# COMPARATIVE EXPERIMENTS WITH FIELD CROPS

# COMPARATIVE EXPERIMENTS WITH FIELD CROPS

G.V. DYKE, M.A. (OXON.),
*Head of the Field Experiments Section,*
*Rothamsted*

LONDON   BUTTERWORTHS

THE BUTTERWORTH GROUP

ENGLAND
Butterworth & Co (Publishers) Ltd
London: 88 Kingsway, WC2B 6AB

AUSTRALIA
Butterworths Pty Ltd
Sydney: 586 Pacific Highway, NSW 2067
Melbourne: 343 Little Collins Street, 3000
Brisbane: 240 Queen Street, 4000

CANADA
Butterworth & Co (Canada) Ltd
Toronto: 14 Curity Avenue, 374

NEW ZEALAND
Butterworths of New Zealand Ltd
Wellington: 26—28 Waring Taylor Street, 1

SOUTH AFRICA
Butterworth & Co (South Africa) (Pty) Ltd
Durban: 152—154 Gale Street

First published 1974

© G. V. Dyke, 1974

ISBN 0 408 70554 X

Text set in 10/11 pt. IBM Press Roman, printed by photolithography, and bound in Great Britain at The Pitman Press, Bath

# PREFACE

This book is written for the man (or woman) who is going to do field experiments for a living, or for fun, or both (I am of this last class). Parts of it are written to help him or her in the interpretation of the results of field experiments. None of it, least of all the chapters on statistics, is written for statisticians. In writing these chapters I have in mind the man so far out in the bush that day-to-day statistics is done by him, or not at all; I have also thought of the man who is diffident about asking his learned statistical colleagues 'what is a degree of freedom?'

Publications that deal with the special problems of experiments with grassland and perennial fruit crops are listed in the bibliography.

If the style is at times light-hearted, I offer no apology; I have enjoyed my work in field experiments and if this is evident, so be it.

This book is loaded with my prejudices; that is why I have so much enjoyed writing it. But I have tried to distinguish my own sometimes controversial opinions by introducing them with 'I think' or some such use of the first person singular. Sentences written in the third person contain only what is in the author's opinion more or less generally accepted doctrine.

I do not want to dress up field experimentation as a science (let's keep 'agronomy' out of this argument) but I think it perhaps deserves recognition as a distinct bit of technology. I have not tried to string together a lot of recipes telling you how to concoct the perfect experiment for every set of circumstances — indeed I have sometimes been at pains *not* to do so. I have tried, by giving examples, to stimulate you to think actively about how you do experiments in *your* circumstances — which are almost certainly different in some material respect from mine. On the other hand where it seems appropriate I have summed up and given a few rules of thumb.

There is a glossary of technical terms on page 176; half the battle of creating a new subject is won when a jargon has been established and I want to do my bit. Each technical term is set in italic where it first appears in the text.

Xagħra, Gozo

G. V. Dyke

21 February 1970

## ACKNOWLEDGEMENTS

I am grateful to many colleagues and ex-colleagues at Rothamsted for goading me into writing this book and helping me with patience and criticism; Harold Garner and Jim McEwen have suffered most. Mohamed Nour lit the fuse by inviting me to lecture on field experiments in the Faculty of Agriculture of the University of Khartoum. Messrs Hunting Technical Services Ltd. gave me two excellent exposures to hot-weather problems. Finally, I am grateful to Rothamsted — a tolerant, liberal, lively community to which I have been privileged to belong all my working life.

# CONTENTS

# Part I

How to Do and Interpret Field Experiments

# Chapter One

# FIELD EXPERIMENTS IN AGRICULTURAL RESEARCH

## 1.1 DEFINITION OF A FIELD EXPERIMENT

It is best to consider that whenever two or more parts of a field receive different *treatments** then a field *experiment* has been started. Most experiments have many more than two differently-treated areas (*'plots'*) but if very many experiments are done to compare the same treatments useful results can be obtained from experiments with very few (even as few as two) plots in each.

Going back a stage further, we should perhaps define a field. (In parts of England and many other countries a 'field' seems an obvious, more or less permanent entity, bounded by walls, hedges or fences, but elsewhere a 'field' is by no means so simple to define.) For the purpose of this book a reasonable working definition is:

A field is a piece of land that has been uniformly cultivated, manured, cropped, etc., in each of the last (say) three years; or failing uniformity, the boundaries between areas with different histories must be known and taken into account in planning an experiment.

In some circumstances three years are not enough to give even a tolerable semblance of uniformity; for example, on soil naturally poor in available phosphorus (P) at Rothamsted heavy applications of superphosphate made in the past still cause large differences in the YIELD

* Words that are included in the Glossary are printed in italic at their first appearance.

of crops after 70 years. A field where such applications have been made on some areas, not on others, cannot be regarded as an acceptable site for an experiment unless the pattern of the old treatments is known and is taken into account in the *design* and *layout* of the present experiment.

My definition does not exclude 'accidental' experiments. If, for example, a field is being ploughed and the work is interrupted by bad weather, it is possible that the areas ploughed before and after the delay will give different yields of the crop grown in the next (or a later) year. You may get useful information from such an 'experiment', crude and fortuitous as it may be. Looking at the other side of the coin, such an accidental application of two or more treatments should be borne in mind and allowed for when laying out an experiment on the site. Methods of doing this will be discussed in Chapter 2.

## 1.2 EXPERIMENTS AND SURVEYS

Surveys can provide excellent information about growers' methods and the yields they obtain but they give little information about the effect of differences in methods. Growers who sow better seed may also give better cultivations and a survey will not reveal how much increase in yield is due to each difference separately. A well-planned series of experiments gives much *information* about the changes in yield that will occur if growers change their methods in certain ways. The functions of surveys and experiments are therefore complementary.

## 1.3 VALIDITY OF EXPERIMENTS

An ideal set of field experiments on one particular subject would be done on many sites and in several seasons. The sites should be a random selection from the whole area of the crop that is being investigated (e.g. all cotton in the Gezira or all wheat on chalk soils in Hampshire). We have to assume that variations of weather between seasons is random. Few people (except perhaps in India) have approached this ideal but we should recognise the limitations of the experiments we do. Growers' methods change as the years pass and the results of a series of experiments done in 1959–1962 may by 1969 be invalid for most of the area of the crop. For instance new varieties of wheat may have stiffer straw and will *respond* to more nitrogen fertiliser than old varieties.

An exception to the above rule: it is legitimate to select special conditions for some experiments. For example, if we wish to compare two forms of fertiliser containing P we do experiments *only* on soils with little available P where P-responses may be expected to be large.

But we *must* state how the sites were chosen in giving the results.

As farming practices improve the experimenter turns his attention from *factors* that have large effects (perhaps doubling the yield) to factors whose effects are relatively small (5% or 10% perhaps). If yields have greatly increased meanwhile the absolute effect (e.g. in kg per hectare) may be about the same. But more refined experiments, with more *replication*, will be needed.

I have deliberately used different units of yield more or less at random in this book – kg per hectare, cwt per acre, and so on. I think an experimenter worth his salt should be ready to use the units most convenient in the immediate circumstances; when in Rome measure yields as the Romans do. I have a respected colleague who records nutrient dressings in grammes per square yard – and why not? (Her balance is graduated in grammes, her measuring tape in feet and inches.)

Field experiments are sometimes done, not to assess the effect on yield of changes in practices, but (for example) to find cheaper ways of achieving the same yields, to compare the efficiency of different systems of draining wet land, or to investigate a matter of theoretical interest such as the availability of P applied in fertiliser many years ago. But in most cases many of the considerations mentioned above still apply.

## 1.4 THE CHOOSING OF TREATMENTS

Although I am not trying in this book to tell you how to choose the treatments to be included in any particular experiment (this would presume a knowledge of your problems and circumstances that I do not have) a few points are worth making.

Most experiments start from some fairly simple question – 'is variety A better than variety B?' or 'will nitrogen (N) fertiliser give a profitable increase in yield?' – and the experimenter thinks of a correspondingly simple set of treatments. If he thinks no more but does an experiment with these treatments (adequately designed and executed) he will get a simple answer, accurate within the limits set by the intrinsic variability of the site used. But the full truth may be more complicated – variety A may yield more than B in the absence of mildew but, because it is exceptionally susceptible to mildew, it may yield much less than B when there is mildew; if no K is applied a single dressing of N may be profitable but doubling the dressing may lessen the profit whereas if adequate K is applied the double dressing of N may be worth while. In such cases the narrow cross-section of the truth that is given by one simple experiment is not enough; the whole, rounded three- (or more-) dimensional truth is what we need. The answer to the original question is a complex one: 'in such and such circumstances, yes, in other circumstances, no'.

A recent example, taken from the Rothamsted Ley-Arable experiment, may be useful. The simple question 'does wheat after three years of *lucerne* yield more than wheat after three years of arable cropping?' has no simple answer. If no N is applied to the wheat the plots that have been in lucerne give more wheat than those that have been in arable cultivation; if plenty of N (say 0.9 cwt N per acre) is applied 'arable' gives more than 'lucerne'. The maximum yield obtainable by varying the amount of N appropriately for each crop-sequence is greater

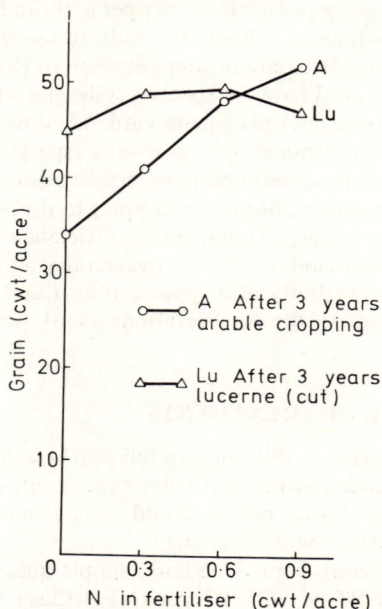

*Figure 1.1* Rothamsted Ley-Arable experiment, Highfield. Mean yields of wheat 1961—63

after 'arable' than after 'lucerne'. Similarly (assuming prices of fertiliser and grain as in 1970) if N is adjusted for maximum profit 'arable' gives more wheat than 'lucerne'. All these statements are needed (and perhaps others I haven't thought of) to answer fully the apparently simple question originally put. The relevant *response-curves* (see *Figure 1.1*) are a convenient means of presenting the facts of the case.


## 1.5  SIMPLE OR COMPLEX EXPERIMENTS?

In general there are two main ways of increasing the usefulness of the answer we get to our question, of increasing the range of circumstances

in which the answer will be applicable. The first is to repeat a simple
experiment at many sites in several seasons, so getting a more or less
fair sample of the variable conditions for which we want the answer
(or answers) to the question. This may be appropriate to the comparison of varieties susceptible and resistant to mildew mentioned
above. The second is to add more treatments to those originally proposed for the experiment. The change will often (but not always)
involve introducing new factors (in the technical sense), that is, including all combinations of the original treatments and some other set of
treatments. For example in testing N at different rates of application it
may be appropriate to use all combinations of these rates and several
rates of K. Or, in the Ley–Arable example, in testing the effects of
different preceding crop-sequences we may need all combinations of
sequences and several rates of N-fertiliser. For examples where the
strictly factorial scheme is not appropriate, see Section 7.5.

This may sound rather daunting and it must be admitted complex
experiments are harder to do, to interpret and to report, than simple
ones. But if real life is complex so also will be much of the work of
investigating and usefully describing it. Some consolation can be found
in the fact that an experiment can often be made factorial (or more
factorial than it is already) without increasing the number or size of plots.

If you are thinking of using 32 plots to compare four varieties with
eight replications each arranged as a *randomised block* of four plots,
you can easily put in a second factor, e.g. N-fertiliser at two rates
('*levels*' in the accepted jargon of statistics). You now have four
replicates which can be laid down as four blocks of eight (this is not
the only possibility – see '*confounding*'). The increase in the size of
each block may cause an increase in *error variance* (because this now has
to take into account differences in fertility between plots that are
further apart) but in most cases that I know this is not serious. But you
can go further ('deeper' might be a better word): why not include a
third factor (K perhaps, or a systemic fungicide to control mildew) also
at two levels? Now you can have two blocks of 16 (each block a
replicate) or by using confounding you can still use blocks of eight plots.
There are several possible designs available but whichever you use you
lose some of the *degrees of freedom* available for estimating the error
variance, so lessening the accuracy of the estimate and (with it) the
likelihood of detecting as 'significant' an effect of any given magnitude.
Put another way, the least significant difference (at a given level of
probability) will be larger than in the simpler design, but the increase is
relatively small. If you use a probability of 0.05 (1 chance in 20) to
define a significant difference the increase is about 3 per cent; if you
use 0.01 (1 chance in 100) it is about 5 per cent. The mesh of the net
through which you strain your results is slightly coarsened, but the

slightly increased risk of missing a 'real' effect, and the slightly decreased precision of the effects estimated from the experiment, are a small price to pay for the widening of the basis of the conclusions due to including the extra factor.

This process can be carried further, to single replicate and fractional-replicate designs. I say no more of these here but end by offering you a rough summary.

By adding in extra factors up to the limit you may lose some accuracy in your estimate of error. But, if there are *interactions* with the added factors you gain in knowledge of a complex mechanism which cannot be simply described; if there are no such interactions, you have most of the value of the degree of replication you would have had with the simpler design.

# Chapter Two

# PLANNING A FIELD EXPERIMENT

## 2.1 CHOICE OF DESIGN

If a statistician is available, consult him about the design of the experiment. He may also suggest extra treatments or modifications of the treatments proposed by the experimenter. In this, and in suggesting the number of replicates needed, he uses his experience of other experiments on similar subjects, and of the methods of interpreting their results.

If no statistician is available, pretend for the time being that *you* are one. Try to get detached temporarily from the muddy practicalities of your programme and consider the logical progression from design to interpretation.

A few words on the functions of a statistician (the word 'biometrician' is used by some people for the chap I am thinking of, leaving 'statistician' for the man who calculates standardised death rates and indices of prices and such like 'observational' statistics). Your statistician's job in considering your proposed experiment is to make your field work as effective as possible for the ends you have in mind. (He may, incidentally, help you to define these ends more clearly.) He will try to give you the design that will give you the maximum amount of relevant information from a given number of plots and he will later help you to extract all the useful information from the results of your experiment. He may, from time to time, throw cold water on your hot head — generally by telling you that so many plots have little or no chance of estimating with useful precision an effect of the magnitude you think likely to occur. This is *not* destructive criticism but is meant to save you wasting your time and effort.

This book is not the place for a full discussion of the designs appropriate to different circumstances but, in case you are your own statistician, here are a few points.

(1) Experiments with many factors, or a few factors at many levels, are most easily arranged in randomised blocks, perhaps with split plots, perhaps with confounding (note, by the way, that these are two aspects of the same thing). For a given number of treatments randomised blocks are easier to lay out on awkward sites than *latin squares* (but see glossary note on unconventional layouts of latin squares).

(2) If the number of treatments is appropriately small, a latin square guards against *fertility trends* better than randomised blocks; a linear trend in any direction is completely eliminated by a latin square whereas randomised blocks deal adequately with trends at right angles to the lines of plots *only* (but other trends can be dealt with less satisfactorily by a modified analysis). Remember in this context that 'fertility trends' include the differential incidence of pests and diseases that arrive by air, often on unpredictable winds.

(3) Where large numbers of treatments of a non-factorial nature (e.g. varieties) are to be compared *lattices* or *lattice squares* are usually the best designs. The uncomfortable restrictions on the numbers of treatments allowed can be dodged by putting in duplicate plots of some of the treatments we wish to test.

(4) So far we have been rather puritanical and have assumed more-or-less rigorous and full *randomisation*. You will meet situations where this is difficult or impossible. If potatoes are planted by machine, spacing between tubers probably cannot be altered during a run that includes several plots. Or you may be investigating the spacing of open drains at 100 m, 200 m and 400 m and a single replicate may be the limit. The answer is, as always, do your best and recognise the limitations of the design (and of the conclusions based on the results of the experiment). Usually (but not always) any comparison is better than none.

## 2.2  HOW TO RANDOMISE

Most experimental designs require randomisation, e.g. randomised blocks, latin squares, lattices. Randomisation (except perhaps in some complex designs) can be done by the experimenter.

Methods of randomisation:

(1) *Tables of random numbers*
If the number of treatments is 10 or less, use one digit at a time from the table; if there are 11–100 treatments use them in pairs. You can save time in various ways, e.g. with 12 treatments, after two have been chosen, re-number the remaining 10 as 0, 1, . . ., 9 and use one digit at a time. If there are three treatments (A, B, C) you can use three digits to represent each, thus:

> A: 0, 1, 2
> B: 3, 4, 5
> C: 6, 7, 8

but the digit 9 must be ignored.

You can use a similar method when the number of treatments is larger; e.g. for 13 treatments, take pairs of digits (00 to 99), divide by 13 and use the remainder to indicate the chosen treatment, but to ensure a 'fair' ('unbiased') randomisation you must treat the pairs 91, 92, 93, . . ., 99 as failures.

You should *not* use the same part of your table of random numbers repeatedly. Either take a random starting point for each randomisation you need or work steadily through, noting the point at which you finish each randomisation. It's not much extra trouble to be rigorous in one of these ways — and it saves you the unpleasant experience of finding the same pattern of treatments turning up repeatedly like a bad dream. If you get to the end of your table, start again at the beginning working down columns instead of across rows.

(2) *Dice, coins, etc.*
Strictly speaking you should check that your die or coin is unbiased, by doing a large number of throws and making

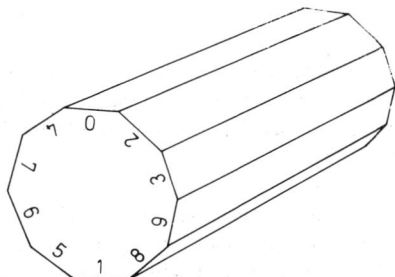

*Figure 2.1* A 10-sides die (With acknowledgements to Messrs Philips)

statistical tests on the results. (The authors of tables of random numbers usually do equivalent tests before publication.) Some of the points noted under (1) above apply to the use of dice – if you have four treatments and a 10-sided die (see *Figure 2.1*) you can allocate two sides to each treatment, using eight in all, but the remaining two sides must be regarded as 'failures'.

If you use coins, and the number of treatments is 2, 4, 8, 16, 32, etc. (any power of 2) the process is fairly simple. For example with 16 treatments the first throw may be used thus:

Heads:   treatments 1, 2, 3, 4, 5, 6, 7, 8
Tails:   treatments 9, 10, 11, 12, 13, 14, 15, 16

The second throw determines which set of four the treatment you are choosing will come from, thus:

|  | *Second throw* | |
| --- | --- | --- |
| *First throw* | *Heads* | *Tails* |
| Heads | 1, 2, 3, 4 | 5, 6, 7, 8 |
| Tails | 9, 10, 11, 12 | 13, 14, 15, 16 |

And so on; four throws will be needed in this case. When eight treatments have been selected the remainder can be renumbered and only three throws will be needed. When the number of treatments is not a power of two you must in effect add in imaginary treatments up to a power of two thus (with five treatments):

| *Treatment* | 1 | 2 | 3 | 4 | 5 | (6) | (7) | (8) |
| --- | --- | --- | --- | --- | --- | --- | --- | --- |
| 1st throw | H | H | H | H | T | T | T | T |
| 2nd throw | H | H | T | T | H | H | T | T |
| 3rd throw | H | T | H | T | H | T | H | T |

The three sequences THT, TTH, TTT are failures.

A short cut is available if you have 3 different coins (one each of 10 p, 5 p, 1 p for instance); throw all together and read them in a definite order. Do *not* throw three similar coins and use the different combinations of heads and tails to define treatments – HHH occurs only one third as often in the long run as HHT (regardless of order).

(3) *Electronic computers*

At least one program has been written (by H. D. Patterson) to produce randomisations; this deals with almost all designs needed for field experiments. It will randomise factorial and non-factorial experiments, it will (if desired) 'restrict' randomisation in the sense of Grundy and Healy[31]. It will deal with split-plot designs and with many types of confounding. The program can randomise factorial experiments with additional treatments. Within reason, you can have your treatments given names or numbers as you please.

The next step, which I hope to see taken soon, is to store the output of the design program in a coded form (perhaps on magnetic tape or disc) and use this as part of the input of the program that will analyse the results of the experiment. This would save the coding of the design (e.g. number of plots per block, what interactions were confounded, etc.) and of the randomisation used. Men being merely human and fallible it will I'm afraid be necessary to build in a way of modifying the coded design to deal with errors at the time of laying-out (e.g. the interchange of treatments on adjacent plots). The object of this refinement, apart from a little elegance for its own sake, is mainly to save a boring job of punching and the errors that may arise in this process.

## 2.3 LAYOUT

The layout (or 'arrangement') of an experiment must be decided in the field. Account must be taken of known or suspected variation of soil characters in the site available including variation in the subsoil invisible on the surface. If in past years cropping or manuring or cultivation has been non-uniform this too must be remembered. Variations above ground that may affect the experimental crop must be foreseen as much as possible, e.g. variations in efficiency of irrigation, the likely arrival of an insect pest from the direction of the prevailing wind. In general it is best to arrange that any known difference in conditions of soil, etc., should coincide with a boundary between blocks (or between rows or columns in a latin or lattice square) so that some blocks are wholly in one set of conditions, the remainder wholly in the other. If there is a change in soil conditions (e.g. the boundary between different crops in a previous year) in such a position that one or more blocks *must* lie across it (or if there is a condition changing gradually in a particular direction, e.g. the effect of shelter due to a line of trees) the harmful effects are minimised if the blocks are arranged so that each plot has the same share of the conditions (e.g. one quarter of its length on the better type of soil) (see *Figure 2.2*).

If machines are to be used for sowing, harvesting, etc., it is an advantage if the layout is such that work on each block (or each row or column in a latin square) can be completed before the next block is tackled; this minimises the disturbance to the results caused by a delay due to mechanical trouble or to weather. In some cases this requirement may conflict with considerations of soil variations. From bitter

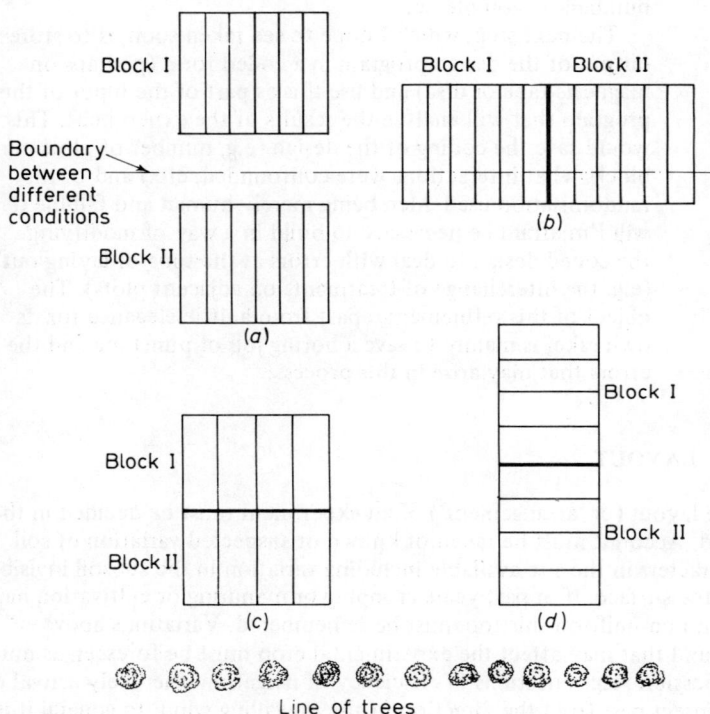

Line of trees

*Figure 2.2* Alternative ways of using a site with simple variability. (a) is preferable to (b). (c) is very much better than (d)

experience of mechanised agriculture in a 'temperate' climate I tend to give precedence to mechanical necessities.

## 2.4  ANALYSIS IN RELATION TO DESIGN

The method of statistical analysis is determined by the design, not the layout; if a randomised block has to be divided, with some plots at a distance from the rest, this does not alter the analysis of the results. A latin square must be analysed as such, even if the layout is not in rows and columns on the field.

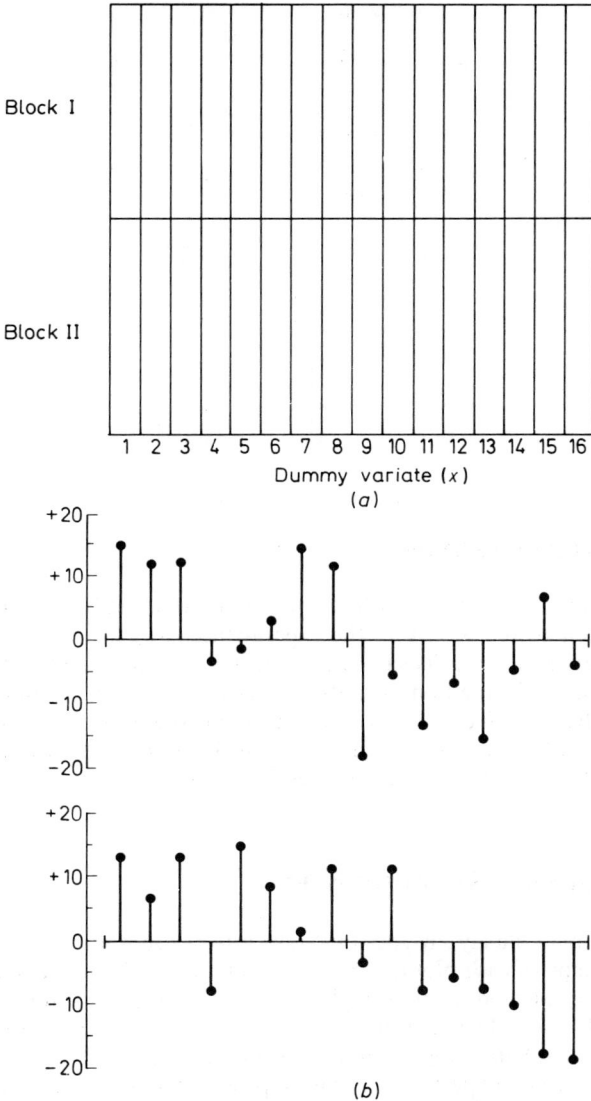

*Figure 2.3* An example of a fertility trend. (a) Layout of the experiment.
(b) Deviations of plot-yields from block means (ignoring treatments),
barley grain (cwt/acre). (c) Scatter diagram to show the correlation
between deviations of yields ($y$-dev) and of position ($x$-dev) after the
elimination of block- and treatment-differences. Mean square for error:
(i) ignoring trend (8 D.F.): 22; (ii) eliminating trend (7 D.F.): 13

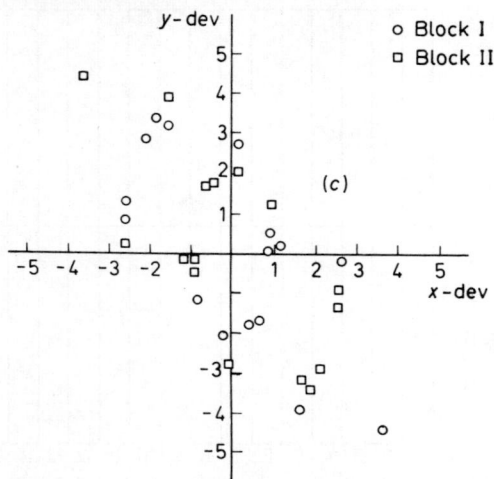

*Figure 2.3 — continued*

## 2.5  FERTILITY TRENDS

Fertility trends (or trends in other conditions affecting the crop) are sometimes detected only after the experimental crop has been harvested; their effects on the results can be investigated, and to some extent eliminated, by the method of analysis of covariance (see Chapter 14). If the analysis is done on a calculating machine (hand or electric) this is arduous but if a suitable program and an electronic computer are available it is easy. For an example see *Figure 2.3.*

## 2.6  HEADLANDS, SURROUNDS, ETC.

When you plan the layout you must take into account, as well as the size and shape of each plot (see Section 2.7), the need for *headlands*, and perhaps sidelands so that implements can be stopped and turned without affecting the experimental crops — remember plough, seed-drills, sprayers, harvesting machinery. Some experiments need a '*surround*' of non-experimental crop to protect them from such things as damage by birds that are worst at the edge of a cropped area. A few need *sidelands* as well as headlands between plots; an example is shown in *Figure 2.4.*

In experiments on the control of airborne pests or diseases there may be *interference* between adjacent plots and it will be necessary to put strips of non-experimental crop between (see *Figure 2.5*). If these are untreated, plots that have received an effective treatment may be affected by reinfection or reinfestation and the treatment will be less effective than it would be if applied to the whole field; if the intervening strips are kept free of the pest or disease (e.g. by repeated spraying)

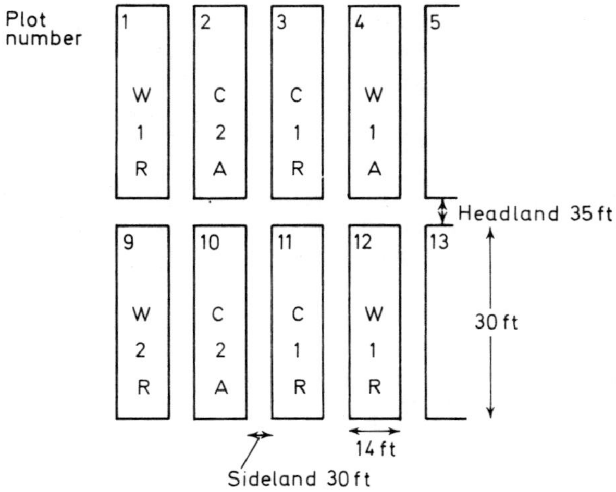

*Figure 2.4* Part of a Rothamsted experiment on beans with headlands and sidelands. *Treatments*. All combinations of: (1) Row spacing close (C), wide (W). (2) Potassium fertiliser at single, double rates (1, 2). (3) Method of application of fertiliser, in bands across the crop-rows (A), in bands near the rows of seed (R). Design: $2^3$ in 4 blocks of 8; randomisation restricted

the damage on the untreated plots may be less than if the whole field had been untreated. In either case the estimated effects are biased. I hope someone will do some experiments in which one set of replicates has treated strips between plots and another set has untreated strips, but as far as I know this has not been done.

## 2.7 SIZE OF PLOT

The size (and shape) of plots is seldom settled entirely by considerations of theoretical efficiency, that is, by the known or supposed relation between plot-size (or shape) and *residual* variance. Points that affect the decision are (roughly in order of importance according to my experience):

(1) the site available;
(2) the *edge-effects* expected;
(3) machinery to be used for sowing, spraying or harvesting;
(4) scarcity of material (e.g. an imported new insecticide) or of manpower (in an experiment involving handwork);
(5) the quantity of 'contaminated' crop to be disposed of in experiments testing materials (e.g. insecticides) whose application to crops that will be eaten by men or animals is not yet approved;
(6) theoretical considerations of *precision.*

This may seem shocking — it is meant to. Here are some illustrations. I would rather use plots smaller than the theoretical optimum (which is seldom very precisely known anyway) than take in land of a different

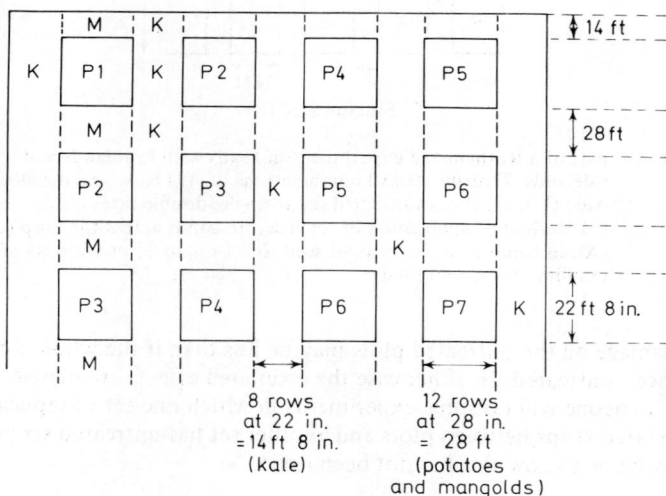

*Figure 2.5*  Based on a Rothamsted experiment with plots of potatoes separated by other crops to minimise 'interference'. P1, P2, etc.: the varieties of potatoes under test for resistance to blight. K, kale; M, mangolds. *Note:* the experimenters would have preferred kale in *all* the inter-plot spaces but this (when full grown) would have precluded inter-row cultivation of the potatoes. Mangolds, which grow less tall, were used in line with the rows of potatoes

soil type, or near to trees, or so near the edge of the field that some
plots would be unduly exposed. In an experiment comparing ploughing
with light cultivation there will inevitably be an open furrow at
boundaries between plots of different treatments. If there is any risk
that one wheel of the seed-drill will run in this furrow the whole drill-
width may be affected and should not be included in the area harvested
for yield. (This is a rather unusual sort of edge-effect.) The widths of
machines obviously influence decisions about plot widths. Spraying,
especially, presents problems as spray booms of commercial machines
are often very wide. Also wheelmarks may cause differences between
plots that are irrelevant to the object of the experiment (see Section 3.2);
or, if wheelmark effects are accepted as part of the treatment effects
they may affect a greater proportion of the harvested area than in a
commercial field. For these reasons a sprayer has been built to spray
small plots with its single wheel running in a narrow path between the
plots (Wilson[49]). In long-term experiments involving different crops a
plot-width has to be found that suits the planting, drilling, spraying, etc.,
of several crops. We have been lucky in being able to use a 7-foot
'module' at Rothamsted for many years. A seed-drill for cereals of 12
spouts spaced at 7 in. (= 7 ft), a potato planter planting two rows at
28 in. (6 rows = 14 ft) and sugar beet sown in rows at 21 in. four rows
(7 ft) at a time all fit well into plots 14 ft wide. We await the advent
of 'metric' machinery with trepidation.

While we are on the subject of plot size, here is a point that arises
occasionally. A site of a particularly awkward shape may suggest the
use of plots of varying sizes (e.g. of fixed width and varying length). I
have seen an experiment in which one of the four replicates had shorter
plots and one had longer plots than the remaining two. This I think was
a mistake. Admitted, all treatments were equally affected, but it seemed
to me that two equally conscientious experimenters could have drawn
different conclusions from the same experiment; A might calculate
means of yields per unit area, B might divide the total produce of each
treatment by the total area of the plots of each treatment. A gives no
extra weight to the larger plots, B weights plots by their area. Final
tables (and *standard errors*) produced by A and B would (in general)
be different.

A slightly different case also arises sometimes. For some reason,
possibly in a long-term experiment, you feel a need to split some plots
while *not* splitting others in the same experiment (or series, if the
experiment is a rotation experiment). There is some risk that edge-
effects will affect yields differently on sub-plots and whole plots. The
only really satisfactory course is to split all plots that are to be compared and give different or identical treatment to the sub-plots of a
plot as appropriate.

## 2.8 PATHS

*Paths* between plots are obviously useful for people inspecting the
experiment, or sampling the growing crop. They may be essential at
harvest (e.g. to allow a combine harvester to stop between successive
plots). They may save marking out several times, e.g. in an experiment
on grass which receives N-fertiliser on several occasions. In a crop grown
in rows there is a distinction between 'down-paths' (i.e. those parallel
to the rows) and 'cross-paths'. The former can conveniently be made by
suitable marking out at the time of sowing or planting. Cross-paths
usually have to be made by destroying the crop later in the season by
cultivation or chemical spraying. Paths can be made just before harvest
by cutting or digging by hand or by a small machine (e.g. for cereals a
front-cut motor-scythe cutting a width of 3 ft). Headlands are needed in
some experiments; they can usually be sown (often by driving the drill
at right angles to the direction of drilling the plots) and narrow paths
can be left unsown or made later to separate the plots from the
headlands.

Paths bring troubles too. One is that weeds often flourish excessively
in paths where there is little competition from the crop. Another is that
paths usually allow the adjacent crop-plants to get more than their fair
share of light and air and of nutrients and water from the soil, so that
their growth and yield may be greater than those of inner plants. They
may, on the other hand, get more than their fair share of damage by
birds (which love paths) and airborne insects which use them,
voluntarily or involuntarily, as routes for invasion.

## 2.9  WHAT IS THE AREA HARVESTED?

In the past many experiments were planned without any allowance for
the effects of paths; the whole area of each plot was harvested for yield
and the yield per acre (or hectare or what you will) calculated by
simple arithmetic. A word here on conventions: I have always taken
the width of a plot (in a crop grown in rows) to be equal to the product

number of rows x spacing between rows

i.e. the crop is considered to occupy a half-space beyond each edge-row.
The length, for a crop grown in 'stations' or 'hills' (e.g. potatoes, cotton)
is similarly calculated, but for crops thickly (and irregularly) sown
such as wheat or grass the length used is taken as exactly the length
between paths or whatever marks are used. Where a machine such as a
combine-harvester or mower is used 'blind' (i.e. without any attempt
to harvest the same set of rows throughout the plot) it is best to base

calculations on the exact width of cut, i.e. the width between the crop-dividers of a combine harvester, or the cutting width of a mower. I have recently, in several such experiments, tested the efficacy of an alternative procedure, viz.,

(1) examine each plot after the yield cut has been taken but before the *discards* are cut;
(2) count the rows cut (or count the uncut rows and do a subtraction sum) at several points along the plot;
(3) use the estimated mean number of rows cut as a basis for calculating yields.

I found this practicable where the rows were 7 or 8 in. apart, but not always possible with rows 4 in. apart (it depends, I think, on the type of cultivation used to cover the seed). To my disappointment this did not increase the precision of the experiments; on the contrary in each case the standard error per plot was slightly increased. I conclude that the sample-points at which rows were counted were too few. Take comfort — you may use this as a pretext for avoiding a rather dull extra job.

If this seems to you like splitting hairs (or straws) think for a moment of the context — experiments in which different inter-row spacings were being compared, at equal rates of seed per acre. The differences in yield were expected to be small, say 5 per cent, perhaps less. The combine harvester used cut a width of 1.5 m. If the combine could be driven with its left-hand divider exactly midway between two rows, and if the rows were exactly evenly spaced (both unlikely contingencies) the number of rows cut would be:

$$8 \text{ at } 7 \text{ in.} = 56 \text{ in.}$$
$$7 \text{ at } 8 \text{ in.} = 56 \text{ in.}$$
$$15 \text{ at } 4 \text{ in.} = 60 \text{ in.}$$

— so here's an artificial difference in yield of 6 per cent ready-made. A row of crop is not a mathematical line and no combine harvester can be driven so accurately. Obviously things will always be rather blurred — judging by small experience, very blurred, which is perhaps a good thing. But if you are working with cotton, or sugar cane or potatoes, and comparing inter-row spacings — think well.

As an alternative to a path of 1 ft width or more, a 'blank' row can be made by blocking the delivery of seed to one spout of the drill. This is commonly done in cereals sown in rows 5–8 in. apart. This gives a clearly seen division between the area to be taken for yield and the adjacent discard. (Discard areas of neighbouring plots can run together, or of course, a path can be left between them.) Weeds do not flourish in the unsown space and edge-effects are lessened though not eliminated. If the soil is dry a nimble man with not-too-big boots can walk in the blank row with little or no damage to the crop. If the crop is lodged a man can walk up the blank row before harvest and part the crop like hair (but using a staff not a comb) so that the combine cuts no more and no less than it should. Some experimenters in calculating yields make an allowance for the better growth of the extreme rows of the yield areas (Widdowson[48]), some do not. If you use this 'blank-row technique' I suggest you investigate the amount of allowance needed in your particular circumstances.

### 2.10  PARTICULAR WAYS OF AVOIDING EDGE-EFFECTS

Two methods of eliminating edge-effects at the ends of plots of potatoes are worth a mention; there may be other crops to which the same ideas can be applied.

(1) Potatoes are planted by normal farm methods (hand or machine); plots, which may be marked out before or after planting (e.g. for fertiliser experiments or for post-planting spray experiments, respectively), are arranged to coincide with definite sets of rows but no attempt is made to start or stop planting exactly at the ends of plots, the whole length including discards being planted uniformly. At harvest an elevator-digger is used. This lifts the potatoes, one or two rows at a time and drops them back on the soil, whence they are picked by hand. The digger like the planter runs steadily from end to end of the planted area without stopping. The yield of each plot is estimated by weighing the potatoes left by the digger lying on a length well within the whole plot area. For instance if the treated plot is 60 ft long yields may be taken from the central 40 ft. The potatoes to be weighed are

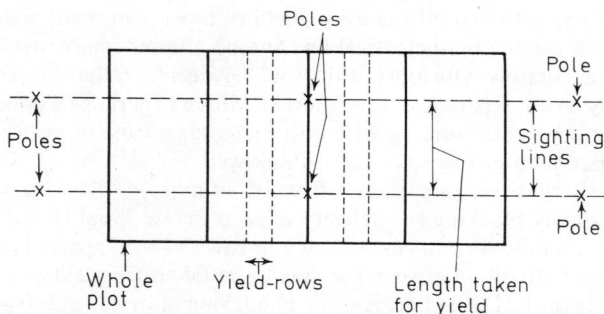

Figure 2.6  A method of demarcating the length to be taken for yield

marked off by putting temporary markers into each row of lifted potatoes immediately after the machine has passed. This can be done by sighting across the experiment using two or three rows of sighting poles, put in before lifting starts (see *Figure 2.6*). The discard length at each end, of 10 ft or more, ensures that virtually no tubers from an adjacent plot will be carried into the yield area. If you use some such method you will be wise to do a few preliminary trials, e.g. note where the first potatoes are deposited when the digger starts on a row; better still, insert some distinctively coloured tubers in a row and note where they are dropped and how much they are mixed with those from neighbouring stations. (You can plant seed of a coloured variety or you can hand-lift the tubers at a station, stain them and replace them.)

(2) Pursuing a line of thought suggested in the last paragraph, you may like to mark your plots at planting time by inserting setts of a distinctive colour in the rows at planting; four seems about the best number. If the whole crop is planted by hand this is easy; if planting is by machine there are two possibilities:

(*a*) Plant uniformly; as soon as possible afterwards dig up setts in discard lengths of row and replace them with coloured setts.

(*b*) With semi-automatic planters where the operators pick setts from boxes and place them in cups you can arrange the setts in narrow shelves within the boxes and put different varieties on successive shelves. This method has been used very effectively in complex experiments where the treatments are combinations of source of seed, variety and chemical treatment of seed; a typical planting sequence in one row might be:

4 setts marker variety
12 setts treatment A
4 setts marker variety
12 setts treatment B
4 setts marker variety

and so on.

If the planter does adjacent rows in the same direction the planting can easily be made to correspond within about one foot (i.e. about unit spacing of setts). If (as is normal in farming practice) the planter does successive runs in opposite directions it is possible, with a few preliminary runs, to register adjacent rows within about 2 ft.

Chapter Three

# SPECIAL CONSIDERATIONS IN PLANNING CERTAIN TYPES OF EXPERIMENT

## 3.1 HOW MANY 'NIL' PLOTS?

Of course the answer depends on circumstances; some experiments need a lot of 'nil' or 'untreated' plots. But I think some wrong decisions are made in this minor part of the planning of experiments. So I give you a few points to consider when you face this question.

First, you can do perfectly good experiments *without* 'nil' plots. Tests of rates of N-fertiliser in experiments comparing crop-rotations are sometimes done without any plots (or sub-plots) without N. If the intention is to investigate the response curve near the point of maximum yield and it is reasonably certain that the test-crop, if grown without N on some or all of the rotations, will yield much less than the maximum, there may be no need for plots without N. But the contrary may be true; you may be interested in assessing the differences between rotations:

(1) in terms of N made available to the test-crop;
(2) in terms of other differences due to preceding cropping, e.g. nematodes or soil structure.

In this case plots without N may help in disentangling (1) and (2).

If you want to compare a new fungicide with Bordeaux mixture on vines you may believe that no grower will ever grow vines without applying a fungicide and so untreated plots are of no interest; the new

24

fungicide will stand or fall by its comparison with the old and differences from 'nil' are not needed. *You* may, but I'm not sure *I* would. Taking this a stage further, you might argue that the presence of untreated plots in the experiment will imply that the treated plots suffer an exceptional concentration of spores (or insects or whatever you're trying to control) and that this will lessen the validity of your experiment. Even worse, you may fear that plots near to 'nil' plots will catch more spores or insects than those further away, an effect known as interference.

So much for 'none' as an answer to the question at the heading of this section.

The next obvious answer is 'one per block' or for latin squares, 'one per row (and of course, one per column)'. A factorial experiment with confounding is an exception to this; if such an experiment includes each factor at rate zero (it needn't) the normal design gives one untreated plot per replicate and each replicate occupies several blocks. The last sentence assumes blithely that all the factors are in effect materials to be put on plots. But some factors may be different, e.g. times of application. In an experiment designed to test three different N-fertilisers (A, B and C) each at rates of 1, 2 and 3 cwt N per acre we have several choices:

(1) omit nil plots and have 9 treatments, represented in a convenient shorthand as

$$(A \; v. \; B \; v. \; C) \times (1 \; v. \; 2 \; v. \; 3)$$

(2) include one 'nil' plot per replicate:

$$(A \; v. \; B \; v. \; C) \times (1 \; v. \; 2 \; v. \; 3) + 0$$

(3) include 3 'nil' plots per replicate.

This last can be thought of as

$$(0 \; v. \; A \; v. \; B \; v. \; C) \times (1 \; v. \; 2 \; v. \; 3)$$
or
$$(A \; v. \; B \; v. \; C) \times (0 \; v. \; 1 \; v. \; 2 \; v. \; 3)$$

with '*dummy*' treatments. In this case the two 'formal' schemes lead to the same set of treatments but this is not always so. For example if the number of materials does not equal the number of rates the two schemes give different numbers of 'nil' plots per replicate. (The meaning of the word 'replicate' is getting just a little bit blurred, by the way.) In particular if confounding is to be used the choice of which factor is to include the zero treatment is important. For example, if a third factor (say P) is involved the $3 \times 3 \times 3$ scheme: $(0\ v.\ A\ v.\ B) \times (N1\ v.\ N2\ v.\ N3) \times (P0\ v.\ P1\ v.\ P2)$ may be easier to handle than the $2 \times 4 \times 3$ scheme: $(A\ v.\ B) \times (N0 \times N1\ v.\ N2\ v.\ N3) \times (P0\ v.\ P1\ v.\ P2)$.

Setting aside cases like these in which factorial structure is important, we come to run-of-the-mill experiments in which we have a set of factorial or non-factorial treatments, confounding is not involved, and we have decided that 'nil' plots will be included. Here are a few points, first for having multiple 'nil' plots in each block.

(1) In any long-term experiment they give the chance for after-thoughts — see Section 7.5.

(2) In a 'rates x types' experiment (e.g. rates of different types of N-fertiliser) you may expect to present mean yields for 'rates' averaged over 'types' (and vice versa) and it's convenient if each mean has the same degree of replication (and therefore, probably, the same standard error — but see Section 11.7).

(3) You may be working with a material that is very scarce or very dear and you want to get the maximum information from a minimum number of treated plots. You may think the standard error of the comparison 'nil' v. 'treated' will be lessened by the extra replication of the 'nil' (despite the increase in the block size).

(4) You may think you would like the extra degrees of freedom for 'error' that come from the comparison of like-treated 'nil' plots within a block — but see Section 11.7 for dangers here.

Now for some thoughts on the other side. The chief one is this: your experiment is often done mainly to compare A with B and the comparisons $(0\ v.\ A)$ and $(0\ v.\ B)$ are of secondary interest. Remember that differences between effects (in this sense) are just differences between treatments:

$$(0\ v.\ A) - (0\ v.\ B) = (A\ v.\ B)$$

But if you are interested in comparing the efficiencies of A and B (they may be two forms of P-fertiliser), then 'nil' plots are essential; you are interested now in response curves and you need the part of the curves near to the zero level.

Finally, if you are short of space (or of time to harvest plots) you may wish for the maximum replication of the treatments even to the extent of sacrificing some of the 'nils'.

## 3.2  COMPARING SPRAY TREATMENTS

In experiments testing chemical sprays (e.g. insecticides, fungicides, growth regulators) it is usual to include unsprayed plots for comparison. For one reason why these plots should *not* be called '*control*' plots, see glossary entry. For other reasons, read on. There may of course be several different chemicals in one experiment, perhaps applied at different rates and different times, but, for simplicity, I deal with one chemical treatment only. I assume the treatment is applied by a tractor-sprayer and that this involves some damage to the crop by wheelmarks. If, instead, a knapsack sprayer is used the operator's boots will do some damage; there is no essential difference. Although plots can be sprayed without wheelmarks (see Section 2.7) and field crops can be sprayed from the air, for many purposes wheelmarks may be considered an inseparable part of the treatment whose effect must be duly allowed for.

The usual comparison is between two treatments:

(1)  Chemical applied in water, wheelmarks on plot.
(2)  No chemical, no water, no wheelmarks.

But we should remember there are four other treatments some or all of which may be worth including:

(3)  Chemical in water, no wheelmarks (by use of sprayers with offset boom, or perhaps by separate harvest of undamaged area of normally-sprayed plot).
(4)  Water alone, no wheelmarks.
(5)  Water alone, with wheelmarks.
(6)  Wheelmarks alone.

Few experimenters would think it worth while to use six treatments to test one chemical at one rate applied at one time, but when several different chemicals are to be tested (perhaps at several rates each) the addition of extra 'untreated' plots [some or all of (4), (5) and (6) above] may be worth while.


## 3.3  COMPARING SOWING (OR PLANTING) DATES

In an experiment comparing dates of sowing or planting it may be necessary to arrange the plots in such a way that each one can be ploughed and cultivated separately; otherwise late-sown plots may suffer because cultivations were done too soon and (perhaps) weeds have grown in the meanwhile.

## 3.4 CULTIVATIONS

In an experiment comparing different cultivations (e.g. ploughing *v*.
light cultivations for winter wheat after potatoes) different plots may
be ready for sowing on different dates. The experimenter must choose
either to:

(1)  sow all plots on the date when the latest ones are ready;
(2)  sow plots of each treatment separately, when ready;
(3)  have two or more different sowing dates incorporated in the
experiment as an additional factor.

(3) is the best (but probably makes the experiment bigger and harder to
do). It allows you to assess separately the direct and indirect effects of
the different cultivations.

## 3.5 TREATMENTS THAT AFFECT RIPENING

Different varieties may ripen on different dates and in a variety trial
you may need a layout that allows each plot to be harvested separately.

## 3.6 COMPARING SOIL FUMIGANTS

In experiments on soil fumigants (nematicides or partial sterilants — all
these classes overlap) some chemicals are injected by hand-operated or
tractor-mounted machine, some are watered on to the soil, some are
applied as powders and mixed with the soil by rotary cultivation. We
may need various 'untreated' plots:

(1)  Soil disturbed by injecter (no chemical injected).
(2)  Water applied, no chemical.
(3)  Rotary cultivated, no chemical,
as well as
(4)  no chemical, no water, no injection, no rotary cultivation
but (as in the case of spraying experiments considered above) most
experimenters are ready to assume that at least some of these treatments will give no measurable effect and so can be omitted.

## 3.7 EXPERIMENTS ON IRRIGATED CROPS

Experiments on irrigated crops present one set of problems if the
irrigation is *basal*, another set if it is a factor being tested. Different
layouts are needed for *basin* irrigation and for overhead irrigation.

(1) Basal irrigation by overhead oscillating spray line or rotating sprinkler. What little experience I have suggests that the amount of water varies greatly from place to place and wind obviously affects the performance of a single spray-line or line of sprinklers. Your layout should take these things into account — for example you might run one spray-line (or two) across each plot of one block — or you might have a spray-line lengthwise between plots 1 and 2, another between 3 and 4 and so on — preferably with an extra one a plot-width outside the experiment near plot 1 and at the other end if the number of plots in the row is even. *Figure 3.1* shows an experiment in which irrigation by sprinklers was confounded with blocks.

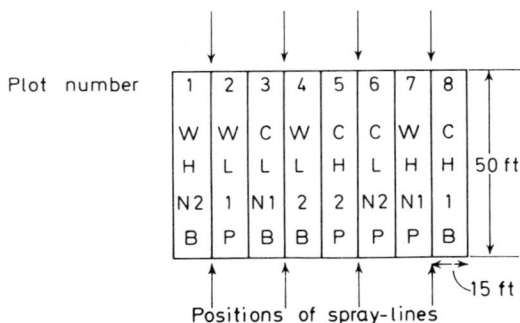

*Figure 3.1*  A Rothamsted experiment on beans (*Vicia faba*) (one block shown; there were eight blocks). *Treatments.* All combinations of: (1) Row spacing: close (C), wide (W). (2) Seed rate; light (L), heavy (H). (3) Fertiliser at 2 rates without N (1, 2), and 2 rates with N (N1, N2) (4) Fertiliser broadcast (B) or band-placed (P). (5) Irrigated (as block shown), not irrigated. *Design:* Single replicate of 4 × 2⁴ with irrigation factor applied to groups of 4 blocks; restricted randomisation of treatments within blocks

(2) Basal irrigation by flow into basins. You have to decide first whether by their nature the treatments must be in separate basins, e.g. soluble fertiliser applied on the surface of the soil may be carried from plot to plot in irrigation water unless each plot is a separate basin. You may still be able to apply another factor (e.g. varieties of crop) on subplots *without* making separate basins. If there is no reason from the nature of the treatments why the plots must be separate basins then you have to decide whether irrigation will be more nearly uniform if separate basins are made for each plot, for each block, each replicate or just one basin for the whole experiment. The decision depends on how nearly level the site is and on the skill of the workers available in applying equal depths of water to similar basins. If each basin contains

more than one plot but not a whole block (e.g. a block of 8 plots in 2 basins of 4) I think the design should take this into account by having sub-blocks coinciding with basins. *Figure 3.2* shows a layout with one

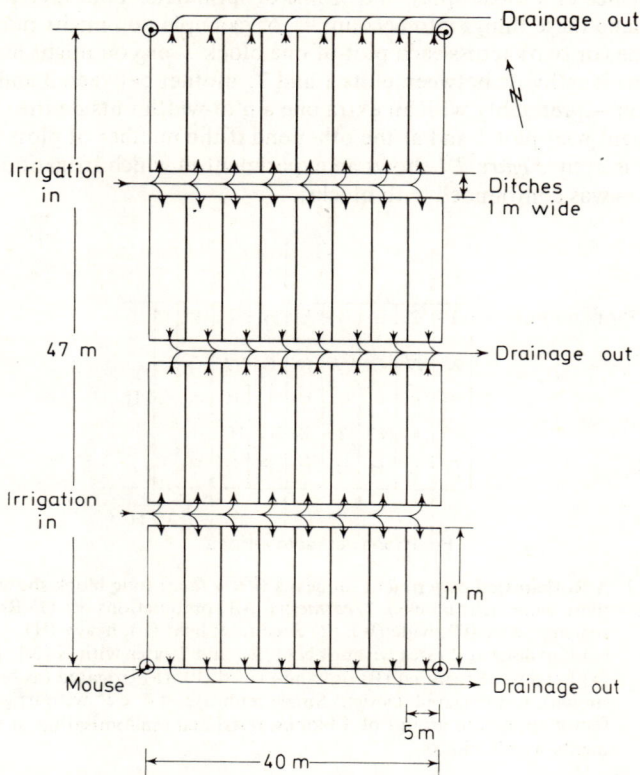

*Figure 3.2* Layout of an experiment on paddy (flooded rice) with flowing irrigation-water. ⊙ indicates a 'mouse'. All measurements to centre-lines of banks (bunds). Cropped area per plot 4 m x 10 m = 40 m²

plot per basin. The crop was paddy rice and irrigation was by continuously moving water; drains were included in the layout.

A useful alternative way of laying out experiments on paddy rice is given by Hauser[4]. He accepts the pattern of *bunds* already existing (or planned by the grower using his skill and local knowledge of differences in level). Each basin (or 'bund') is used as one plot, with the bund as boundary. This means that plots are not exactly rectangular and their areas are not all equal. Hauser (who is writing about fertiliser experiments) therefore calculates the appropriate quantity of each fertiliser for each plot using the plot-area rounded to the nearest 5 m². Edges are discarded at harvest and yields are recorded from equal areas from all plots. This avoids the difficulties of interpretation that are caused by unequal harvested areas (see Section 2.7). Note that Hauser tacitly assumes the necessity of having different fertiliser treatments separated by a bund; he would not allow water to flow from one treatment to another. Gasser[30], however, working in British Guiana (now Guyana), believed there was no harm in having different fertiliser treatments as basin-fellows.

(3) Treatment irrigation by overhead application. Rotating sprinklers (except perhaps the 'logarithmic' type consisting of a rotating boom with nozzles whose spacing varies inversely with the distance along the boom from the pivot) seem to me almost unusable except for relatively crude experiments comparing 'some water' with 'no water'. Perhaps if sprinklers with a small radius of application are moved frequently within a relatively large plot a reasonable experiment could be done. With oscillating spray-lines fairly even application can be made (and water can easily be metered) to plots of perhaps 10 m² or bigger but there are substantial areas at the edges that receive less than the due rate of irrigation and these should be discarded.

(4) Treatment irrigation in basins. This is great fun – as good as making castles and moats on a sandy beach. Two things need attention (apart from ensuring that the bunds do not leak!).

(a) Try to apply equal amounts to plots receiving the same treatment – this may be done by measuring the flow in the ditch with a measuring weir (e.g. a V-notch) or some other metering device and allowing the appropriate time of flow; for treatments that should receive more or less water than was applied to the first treatment the time can be modified accordingly.

(b) Try to apply the same amount in successive irrigations (unless there is reason to do otherwise). This probably applies specially where different frequencies of irrigation are being compared. I did this once on a soil which accepted water slowly by filling each basin the first time equally (by judgment) and immediately hammering in pegs to indicate the water level. Actually it was more complicated – I wished to test, at the later irrigations, differing quantities of water, so plots of different treatments were filled to a certain level below or above the level of the first irrigation. I am not sure this was the best way of doing this experiment.

There are other ways of watering crops (of which I know little). For example long-furrow irrigation, which seems difficult to fit into any scheme of treatment irrigation. Plastic hosepipes with pinholes, though very different from most growers' practice, offer a chance to do tests of irrigation on small plots. A hose held in the hand, and even a humble watering can, have their place too.

## 3.8 EXPERIMENTS ON TECHNIQUE

Experiments designed to compare different ways of doing experiments are discussed in Section 9.6.

# Chapter Four

# MARKING OUT, SOWING, COUNTING, SCORING

## 4.1 MARKING OUT

Your design is settled, your randomisation done, your layout, by a process of compromise between what you want and what you have, is decided and your *sketch* is drawn. Now to mark out the site for the application of treatments, the sowing of seed, or the planting of tubers or cuttings. Or you may be marking out for an experiment on an existing crop, e.g. grass.

Start by establishing a base-line. This sounds grand and is intended to give you an early feeling of confidence but (often) there's no real distinction between the base-line and the other edges of the experimental site. But you've got to start somewhere and where you start can be conveniently called your base-line. Where you put it depends on many things. Obviously you try to place it so that the site as later marked out avoids the pitfalls (metaphorical ones *only* I hope) of the field – tree-shadows, open furrows and gathering ridges of fields ploughed in 'lands', boundaries between different crops or manures applied in recent years, and so on. You will also bear in mind whether you have the whole field to play with or whether other experimental sites, in the same or later years, must be available. You will have to take into account the direction of crop-rows if you are marking out after sowing; when marking out before sowing you should consider which way the rows will run in the whole field and conform to this if possible. Your baseline may be parallel to the rows or at right-angles to them. If several experiments are to be in the same field it is often convenient to use a common base-line for all or most of them. Generally it is best to establish as base-line the longest line in the final layout of the experiment(s) in the field.

Having established your base-line, firmly marked with two stakes or sighting poles, start on the first experiment. Mark out the rectangle which will contain the whole experiment (occasionally a more complicated shape, e.g. if one block has to be a little to one side to avoid a hazard such as the shade of a tree). Work round the rectangle, marking the appropriate lengths at right angles; the end of the last side should coincide with the starting-point. In practice it seldom does exactly. At this stage you make minor adjustments to the angles (not the lengths), and establish the corners once and for all (see *Figure 4.1*). An error across the width of a block of plots of 1 per cent or less is not worth correcting. Intermediate points on the four sides are marked next — plot boundaries, headlands, etc. When you have marked the intermediate points in one side check the measurement to the corner at the far end. If there is a discrepancy adjust all the sections (e.g. plot-widths)

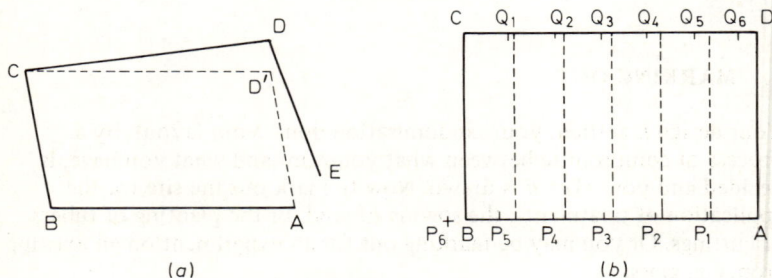

(a)                          (b)

*Figure 4.1*  Stages in marking out an experiment. Inaccuracies exaggerated for clarity! (a) Stage 1. ABCD = first try, AE = discrepancy, ABCD' = corrected 'rectangle' used (CD' = CD, DA' = DE); (b) Stage 2. Dotted lines show plot boundaries used $P_1, P_2, \ldots, P_6$ and $Q_1, Q_2, \ldots, Q_6$ are marks put in at first try. ($P_6$B and $Q_6$D are discrepancies)

to conform (see *Figure 4.1*). Points that are needed inside the rectangle can be marked by putting down lines (of string or thin rope) from edge to edge of the experiment in both directions, or by putting down lines in one direction and 'sighting-in' in the other; sighting-in in both directions is usually frustrating to the three people involved.

## 4.2  SIGHTING-IN

'Sighting-in' can be done in two ways (see *Figure 4.2*). If a third point C is required on the line joining two fixed points A and B one man standing at (say) A directs a helper to move a marker until it and those at A and B are in alignment. (This sounds, and is, simple, but I've met

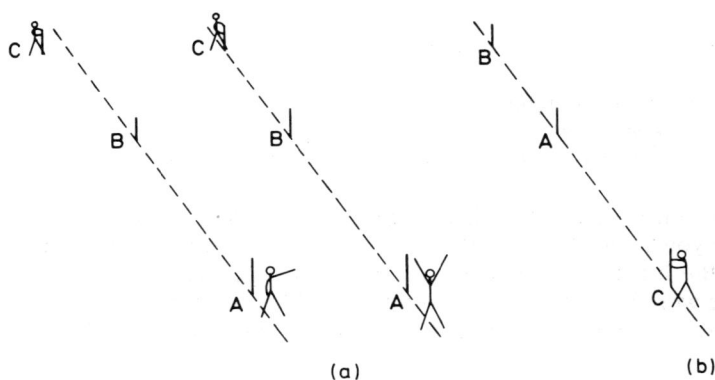

*Figure 4.2* Sighting-in, and sighting-back. (a) sighting in (duet); (b) sighting back (solo)

people who seemed unable to grasp the idea of alignment — so check on your helpers — tactfully.) If the third point required lies beyond (say) A it can be settled by one man 'sighting-back', that is moving his own marker sideways until he sees it, and those at A and B in alignment.

## 4.3 PARALLELOGRAM-SHAPED PLOTS

In most experiments the plots are rectangles or squares but this is not essential and (as in the case of the barley experiment started at Rothamsted in 1852) if the site available is roughly a parallelogram with adjacent sides at an angle of say 80° you may consider making each plot a parallelogram so that the experiment conforms to the site and uses it fully. Difficulties that arise are:

(1) Plot areas are not simply 'length x breadth' but the correction required (multiplication by the sine of the angle between adjacent sides) is the same for all plots and (unless the angle differs much from 90°) makes very little difference and may perhaps be ignored.

(2) Machines with a great width of working (e.g. sprayers, combine harvesters) will need wider-than-usual paths, or headlands, to stop and start or to turn between plots.

## 4.4 PRACTICAL POINTS IN MARKING OUT

(1) Right angles can be set out in many ways:

>  (*a*)  by an *optical square*;
>  (*b*)  by a *crosshead*;
>  (*c*)  by setting out a triangle with sides in the ratio 3:4:5.

The optical square is small and light, is held in the hand (*over* the corner you are working from) and you see both directions at the same time. But it takes a little time to get used to an optical square.

The working part of an optical square is either one prism or two pre-set mirrors (see *Figure 4.3*). In the mirror type it is possible to dislodge

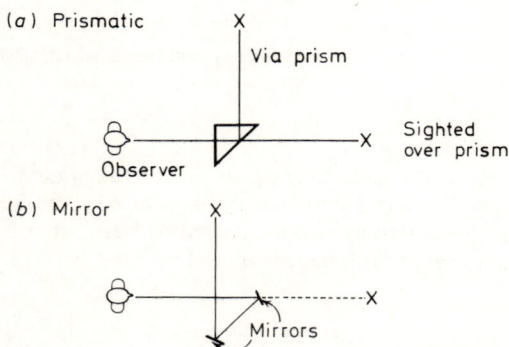

*Figure 4.3* Optical squares

one or both of the mirrors (e.g. by dropping the square on a hard road) and you may then get a false angle before you realise anything is wrong. The prismatic type is smaller (small enough to be forgotten in a man's pocket) and cannot fail unless the prism is broken or dislodged, which would happen only with the most callous treatment.

Double optical squares are available, with two prisms, one above the other; with one of these you see to left and right in the two prisms, and ahead over the top or between the prisms.

A crosshead is heavier and bulkier than an optical square but its main disadvantage is that it has to be mounted on a staff stuck in the soil and it (or the staff) twisted till it is aligned correctly, then left untouched while the user moves round to sight in the other direction. Crossheads are robust, and you can easily see if they are seriously damaged.

The 3:4:5 method requires three people, a long, accurate measuring tape and much care. The best I can say for it is that it is a good one to fall back on when you have driven 20 miles to your site and find you have forgotten to bring your square or crosshead.
When you set out a right angle it is best to have a mark at a good distance — further than the length you need. It is easier (but less accurate) to put a mark closer and 'sight back' when you measure the length. For other angles (and for that matter, right angles) you can use either a sextant or a prismatic compass.

(2) Lengths can be set out using either chains which are heavy and may stretch slightly with long use, or tapes of linen, steel or plastic. Tapes cannot be pulled straight in a strong side wind and you have to be adroit in laying them down on the soil with a sideways movement. Tapes are harder to clean than chains and linen or steel tapes suffer if rolled up wet. All tapes tend to go into kinks and steel tapes can cut a bare leg. Most tapes have both metric and imperial markings (an advantage over chains). Chains can be obtained with each link equal to 1 ft or 0.5 m (or, of course, 1 link, i.e. 7.92 in. so that 100 links = 66 ft — this is the basic reason why, until recently, many plots at Rothamsted were equal in length to a cricket pitch).

(3) The corners of the experiment (and any other points likely to be needed often for sighting) should be marked with substantial poles or stakes. The intermediate points can be marked with small pegs of wood or metal. 'Surveyors arrows' are suitable and a man can carry 50 of these easily. When setting out long distances they are the best markers to put in temporarily at the end of each length of the tape or chain. You will lose less of your arrows if you tie a bit of red string to the loop of each one.
For marking corners at distances of not more than about 100 yards sharpened wooden pegs about 3 ft long, 3 in. wide and 1 in. thick are useful. For work in a tall crop or on uneven ground over long distances, you may need surveyors' poles 6 ft or so long. These have sharp iron or steel points and can be jabbed into hard ground better than wooden pegs. The best colour in most conditions is white; a combination of red and white as often used on surveyors' poles is also good.

(4) Walk just *outside* the boundary of your site when marking out, not inside.

(5) Remember that you will have to find the corners of your experiment accurately later on, when cultivations, grazing animals, passers-by or termites will have eliminated most or all of the pegs. For an experiment lasting only one year you may be able to mark the corners by a combination of down-paths and cross-drilled headlands but in most cases something tangible in the soil is needed. 'Mice' (see *Figure 4.4*) are easy to place — their burrows can be made with a 4-in. soil auger.

'Mice' have some disadvantages — their wire tails upset the digestions of mowers and combine harvesters, looking for a single wire in a cereal field is horribly like looking for a needle in a haystack, and, if the local people prize galvanised wire, 'mice rustlers' may scour the fields. Heaps

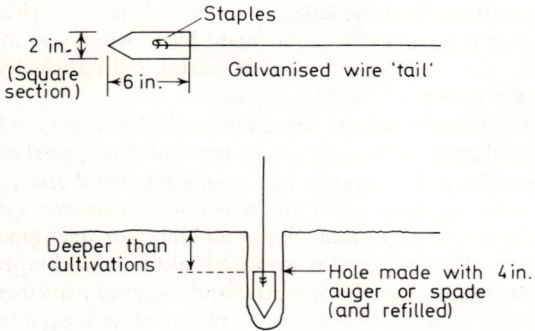

*Figure 4.4* A 'mouse' and its burrow

of soil are used as markers of crop-strips in some areas and a 'mouse' concealed in such a heap may escape the rustlers.

For experiments that last several years something more permanent (and without the 'tail') may be needed. At Rothamsted blocks of oak about 4 in. square by 6 in. deep, heavily creosoted and put in the subsoil below plough depth last 5—10 years; their upper faces are found when needed by digging the topsoil away carefully. Another method used successfully involves putting in the subsoil a suitably massive metal object which can be located by a 'mine-detector' without digging (Anon.[13]; Jameson and Mansbridge[35]).

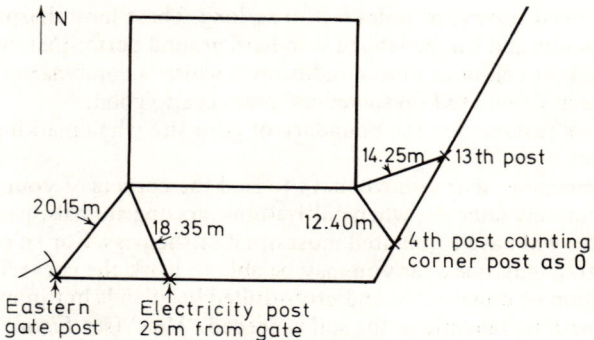

*Figure 4.5* Tie-in measurements

Whether your experiment is to last one year or more you will almost certainly need to take measurements from one or more corners to identifiable points nearby, e.g. buildings, roads, gate-posts, trees, telegraph poles. (But remember, if it is not your land, that most of these things may be moved without advance warning!) It is usually easier to measure to two fixed points and rely on distances only (see *Figure 4.5*) than to measure angles also — but this depends on circumstances. Experimental farms should have permanent markers specially put in at suitable points.

## 4.5  SOWING, PLANTING, ETC.

Methods vary with circumstances but here are a few points.
    If a plot is 2 or more widths of the seed-drill the tractor driver

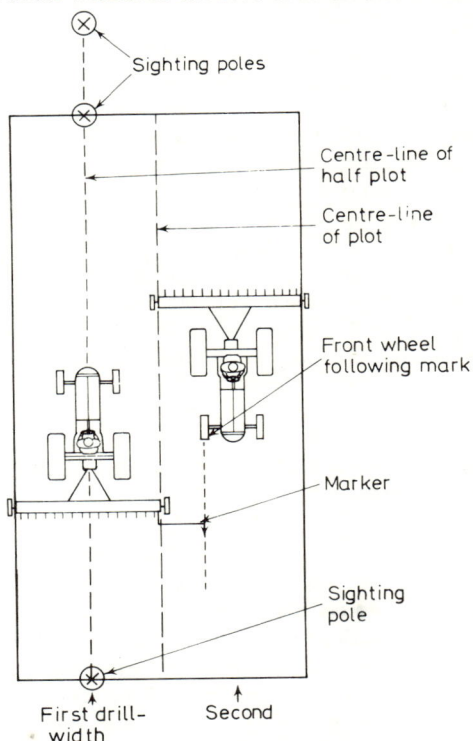

*Figure 4.6*  Setting out a plot for the drill (in this example the plot is 2 drill-widths wide). The 3 sighting poles or pegs are put in temporarily for the driver; they are knocked over by the tractor

usually does the second and later widths by driving with one wheel on the mark in the soil made by the device attached to the drill for this purpose. To keep him straight (and in the right place) on the first run, put conspicuous markers on the centre-line of the drill-width, one at each end of the plot and one about 10 m beyond the far end (see *Figure 4.6*). The driver drives over the first marker and then steers so as to keep the other two in line till he completes the length.

When drilling an experiment in which several varieties are compared it is usually convenient to drill all the plots of one variety before clearing out the seed-hopper and starting with the next variety. There are a few drills, however, designed so that, whatever quantity of seed is put in, the whole lot is sown on a predetermined length. With such a drill it may be better to do the plots in systematic order, changing variety at plot boundaries as required. Plots may be separated by relatively narrow cross-paths instead of the wide headlands needed if the drill has to go from plot to plot of one variety.

If the treatments involve different chemical dressings applied to the seed much the same applies, but it may be necessary to clean out the seed-hopper especially thoroughly. If the experiment includes untreated seed and seed treated with different rates of the same material, and you drill in increasing order from none to the heaviest dressing, it may be unnecessary to clean out between treatments. When different seed-rates are to be compared the procedure chosen will depend on the mechanism of the drill. If the rate of seeding can be changed quickly and reproducibly (e.g. as in changing gear in a car) you can drill the plots systematically. If, by contrast, you need to replace one cog by another, or move a shutter with a continuously variable setting it will probably be better to drill all the plots of one seed-rate before making a change.

## 4.6 CALIBRATION OF SEED-DRILLS, FERTILISER DISTRIBUTORS, ETC.

Whether a machine is to be used for basal or treatment application it is necessary to know beforehand what rate of seed, fertiliser, etc., per acre it will apply, and usually to choose from a range of settings the best one or more for the experiment in hand. For this it may be necessary to do a fresh calibration before each experiment.

Calibration can be done with the machine stationary or moving. In stationary calibration the wheels are jacked up and turned by hand a given number of turns. From the quantity of seed or fertiliser delivered the rate per acre can be calculated. This method cannot easily be applied to a machine driven from the power take-off of a tractor. This arises in calibrating sprayers and necessitates complicated exercises with a stop-watch and pre-arranged engine speeds or tank pressures.

It is generally better to do calibrations on the move. The machine should be run over soil in the same condition as in the experiment. This ensures that the same amount of wheel-slip will occur and also that seed or fertiliser will shake down and settle in the hopper to the same degree. Remember that many machines give different rates of delivery when working up and down on a slope.

If possible do replicate runs at each setting and so get some idea of the accuracy of your calibration. When you are calibrating a seed drill it may be worth while to collect and weigh the seed from different spouts separately. Similarly when calibrating a fertiliser distributor you may be able to collect the fertiliser in a divided tray and get some idea of the variability across the width of the machine. In calibrating sprayers you should collect the liquid delivered by each nozzle separately. Be prepared for some nasty shocks in all these comparisons between apparently similar bits of the same machine.

There is a delay before starting or stopping the mechanism of a drill or distributor takes effect, so the longer the run for calibration the better; this applies to stationary calibration too. When on the plots the drill must be started before the end of the plot, and stopped after the far end. This means that it is *not* easy to calculate the actual rate of seeding by finding out how much seed was used on the experiment.

When different varieties are compared it is usually necessary to calibrate the drill separately for each variety. You have to decide before-hand whether you want to sow equal *weights* or equal *numbers* of seed of the different varieties; you may also need to make allowances if the percentage germination varies between varieties.


## 4.7  APPLICATION OF BASAL MANURES, SPRAYS, ETC.

If a basal fertiliser is to be applied by hand it is better to weigh out the appropriate amount per plot and apply to each plot separately than to do the job by (say) blocks instead of plots; however carefully you do the distribution over a block the individual plots will receive slightly different quantities. On the other hand, if the person doing the spread-ing has bad habits (e.g. putting too much fertiliser near the boundary of the area marked out) he may do less harm if he works on a whole block (or better still a block plus a border all round).

If the fertiliser is to be put on by a machine, arrange for it to work across the plots rather than along their lengths. Then, if the machine's behaviour changes as the container empties the resultant variation will affect many plots more or less equally. This is also an insurance against gross troubles (e.g. the container being filled with the wrong stuff the first time, or the rate being double the intended rate).

Whenever possible a basal operation should be done one block at a time (or, if the design is doubly restricted, e.g. a latin square or lattice square) either one row at a time or one column at a time. This ensures that, if the job has to be done on two or more separate occasions (perhaps because of unexpected rain or the breakdown of a machine) any differences in yield that may result will inflate differences between blocks (for 'blocks' read 'rows or columns' for the designs with double restriction).

If a job done by hand is likely to be done with different degrees of skill by different workers (singling sugar beet is a conspicuous example) you should arrange that all the plots of one block (or row or column) are dealt with by the same man or men; if several men work together each should do an equal share of every plot. The same applies when several men operate machines — every machine does a better job when adjusted and guided by a skilled and conscientious man than otherwise. Most jobs done on an experiment by machines can be done by one man so this point arises rather seldom. It is worth while, however, to discourage your men from resetting machines in the middle of a block; you can easily provide a dummy plot alongside the experiment on which all implements can be tried out and adjusted before a start is made on the experiment proper.

There are exceptions to this rule; if the soil at one end of a block is harder than at the other a machine that works well at the beginning may *have* to be readjusted to give an equal performance at the end. There are cases too in which the treatments themselves cause differences that have to be taken into account. Phosphate fertiliser applied to sugar beet, for example, sometimes greatly speeds early growth and it may be necessary to single beet on P-treated plots before the remainder. The considerations set out above apply equally to harvesting, sampling or making counts, etc., on the plots of an experiment.

To sum up: put all the sources of unwanted variation you know of (or suspect) on blocks or rows or columns.

## 4.8  PLANT COUNTS, SCORES, ETC.

Go and look at your experiment often. If you see no difference between the plots and no irregularity of growth, etc., note this down and note also the state of the crop, of the soil, of any weeds or pests or diseases you see. If you do see differences or irregularities note them too. If the variation does not correspond to plot differences, e.g. if an area has been flooded (or left un-irrigated) make a rough drawing with measurements. If the variation you see corresponds with plot differences make numerical scores, one for each plot. Do this on blank paper or on a previously-prepared sheet with plot numbers entered [not the plan, as a knowledge of the treatment of a plot may predispose you to give it a large (or small) score]. Try to score in whole numbers, say 0–10 or 1–10 or 1–5. It usually pays to look round quickly before starting scoring formally, to assess the range of variation and to decide how many classes you can distinguish. Work by blocks and if you can stand back and score two or three plots from one point it helps to keep your standard constant. If you find a plot in which part of the area is different from the rest there seems no simple alternative to recording a score that is a roughly-weighted average of the whole plot. Note down what your smallest and largest scores mean (e.g. '0 = 2 in. high, 5 = 6 in.

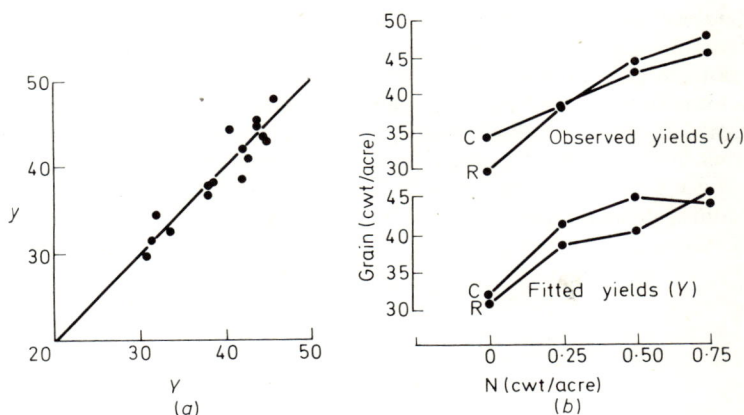

Figure 4.7  An example of the relation between visual scores and yields. The crop was winter wheat, the 16 treatments were the combinations of 4 rates of N applied in spring and 4 types of 1-year ley grown in the preceding season. The plots were scored in May, the scores ranging from 2 to 6½. The regression of yield of grain in cwt/acre $(y)$ on score $(x)$ was estimated from the means for the 16 treatments. The fitted values $(Y)$ were calculated from the regression equation

$$Y = y + b(x - \bar{x})$$

The goodness of the fit is shown in (a). The coefficient of correlation between observed and fitted yields was + 0.94. Another way of illustrating the fit follows (I omit 2 of the 4 types of ley for clarity) [see (b)]. The state of growth of the wheat on 27 May (before the ears had emerged) reflected much of the pattern of the yields recorded on 30 August, for example the greater response to N after ryegrass (R) than after clover (C). A testimonial to excellent scoring by H. V. Garner

high' or '1 = pale 3 = dark green' or '0 = no lodging, 10 = all lodged'). If you are not sure whether the variation you see goes with plots or not (e.g. a condition that occurs in small patches) score it, analyse the figures and then you will have an idea whether it was affected by treatments.

When you can count plants on sample areas or measure their height this will almost certainly be better than making scores for these characters, but scores do take account of the whole plot and can be done quickly and without walking through the crop. I think it is better to score all your experiments (and count none) than to count half and do neither on the remainder. For methods of selecting sample areas for counts, etc., see Chapter 6.

One last point about scores: *use* your scores. They may help you to interpret the mode of action of a treatment or the difference between the effects of the same treatment on different sites (see *Figure 4.7*). They may help you to decide whether or not to accept an exceptional yield (much more or less than the value expected from block- and treatment-effects); if scoring showed that the corresponding plot had an exceptionally good (or bad) appearance without evidence of a 'gross error' (e.g. signs that a cow has been rolling in the plot) you may decide to accept the recorded yield. If, on the other hand, the score for the plot was not exceptional you may feel justified in rejecting the recorded yield and proceed to estimate a 'missing value' (see Section 13.9).

# Chapter Five

# HARVEST

## 5.1 IMPORTANCE OF YIELDS

Most field experiments are run through to harvest and usually the yields recorded are an important part of the results, either for themselves or as a basis for other calculations (e.g. *uptake* of $K$). A really bad mistake in recording yields can often render useless a great deal of work over a period of months or even years. So it is worth being a bit solemn and careful about how the job is done.

## 5.2 WHAT AREA PER PLOT FOR YIELD?

At harvest-time most experimenters discard the crop growing at the edge of each plot. Plants growing near the edge of a plot are often different from those further from the edge. If there is a *fallow* path round each plot the plants at the edge often grow better than those in the middle although occasionally the opposite effect is seen.

In experiments laid out with very narrow paths (or no paths at all) we may be ready to neglect edge-effects of the types just mentioned but in this case the plants at the edge of one plot may have neighbours on one side whose treatment is different from their own. Their growth (above or below ground) may be more or less vigorous and so the edge-plants will suffer more or less competition on that side than plants further from the edge. Whether their yield is increased or decreased depends on the treatment of the neighbouring plot.

If the whole *area* of each plot is harvested and weighed for yield troubles of different sorts can be expected. Where paths are present (and inter-plot effects are small) we may be fairly sure (for example) that all yields are increased by edge-effects; but are they all equally increased or is the increase proportional to yield, or what? We do not know and we cannot tell how our results have been affected. Where there are no paths we may be pretty sure that plots whose yields are small (because of their treatment) will give recorded yields less than the 'true' yields because their edges will have suffered excess competition while plots with bigger yields will yield more than the 'true' figures. The effects of treatments will be overestimated, by unknown amounts.

Having decided what areas must be discarded at the sides and ends of plots the experimenter really has two choices: to harvest the whole of the remaining area or to harvest one or more relatively small sample areas. Sampling will be dealt with later (see Chapter 6) but one thing must be made clear now. If other things were equal (they never are) harvesting whole plots would always be better than harvesting sample areas only.

Yield varies from sample area to sample area within a plot and the more sample areas you take (the limit being the whole plot less discards) the more accurate (in general) the mean results will be. Note that in a particular case by a fluke, the standard error per plot from sample harvest may be less than the standard error from whole-plot harvest. A strictly analogous fluke will illustrate this: a man tosses an unbiased coin 100 times. After 50 tosses he has recorded 25 heads (i.e. the proportion of heads is 0.5, exactly equal to the value expected in a very large number of trials) but after 100 tosses he has 45 heads, a proportion of 0.45 — an estimate that is less accurate although based on a larger sample.

Now some general points about harvesting whole plots (less discards) — some of these apply to sample harvest too.

## 5.3 MARKING OUT THE AREA FOR HARVEST

The area harvested should be marked out in the same way for all plots. Make sure that the area harvested is the same on all plots, especially within a block. (There are a few necessary exceptions; for example in an experiment comparing different harvesting methods you may want plot size as a factor, and in an experiment comparing a crop sown in rows with the same crop sown broadcast you may harvest a chosen number of rows in the former, a cutter-bar width in the latter.)

In any case write down *at the time* the dimensions of the harvested area of each plot as you have defined it at the time. Do *not* do any calculations in the field — or if you do, throw them away and leave just the raw figures (e.g. write '100 ft x 4 rows at 21 in.' if appropriate rather than '100 ft x 7 ft', or '700 ft$^2$'). Don't worry about funny mixtures ('60 ft x 15 rows at 10 cm') — they can be sorted out at leisure later. Don't forget to note any exceptions, e.g. 'plot 51, 50 ft harvested instead of 60 because the drill failed to sow the last 6 ft'.

## 5.4  LABELLING THE PRODUCE

If the produce of each plot is to be weighed on the plot you can leave it there without labels. If the produce of the plots is to be carted to one place for weighing it must be labelled with plot numbers (or perhaps block and treatment numbers). Duplicate labels should be used; produce that goes in bags can have one label inside, one tied on outside. (A convenient form of label perforated for easy tearing and stapled together into blocks of about 20 is illustrated in *Figure 5.1.*) Check beforehand

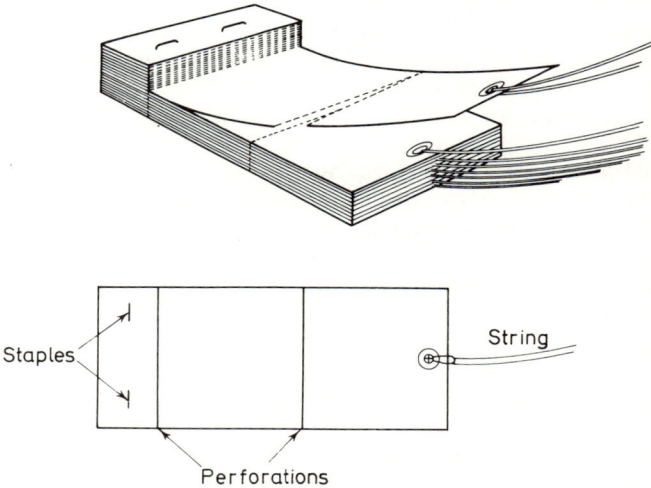

*Figure 5.1*  Perforated labels

that the labels and the ink, pencil or other marker you intend to use can stand sun and water and any other local hazards likely to affect them in the field. Felt-tipped or fibre-tipped markers with non-water-soluble inks (e.g. 'Scripto', 'Pentel') are good.

## 5.5  WEIGHING THE PRODUCE

The produce of all the plots of an experiment should be weighed to the same degree of accuracy (e.g. to the nearest lb). The degree of accuracy needed is a matter for the experimenter to decide. For guidance on the choice of a rounding unit see Appendix A. For the present let us say that a rounding unit of about 1 per cent of the mean plot yield is acceptable for all types of experiments known to me.

Unless you are using direct-recording balances (see Appendix C) have two people reading the pointer of the balance. One writes down his reading and checks it with the figure called out by his companion.

For most types of produce the most convenient and quickest way of weighing is by a spring balance hung from a beam or a tripod (if many weighings are done in one place) or from a mobile jib if the balance is to be moved frequently. Spring balances are extremely durable, accepting knocks, dirt and damp without losing accuracy. But check your balance regularly by hanging on the hook a known weight.

Most spring balances have pointers that can be offset so that a tare is subtracted automatically. If each lot of produce is in a sack of a standard weight (say 1 lb) you adjust the pointer to read zero when an empty sack is on the hook and thereafter you will read net weights directly. A note about tares must accompany the weights (e.g. 'constant tare of 1 lb to be subtracted' or 'no tare' or 'variable tare recorded as $T \ldots$').

A useful device for lifting sacks by the neck, quicker than a running noose of chain, can be attached more or less permanently to the hook of the balance (and its tare can be allowed for by moving the pointer). This is rather like a pair of tongs and once the jaws are applied fairly to the sack it grips well, using pressure in proportion to the weight of the sack.

## 5.6  TAKING SAMPLES FROM THE HARVESTED PRODUCE

In many experiments samples of the harvested crop are taken; for example samples of grain for the determination of the percentage dry matter, samples of sugar beet roots for the determination of percentage sugar. It is best to take these at the time of weighing the produce so that if any changes are taking place (for example, if the harvested grain is losing moisture to the air) the samples truly represent the material weighed. It is sometimes convenient to include the samples in the plot weighings, sometimes it is better to weigh them separately and add the weights later. You must take care to get a representative sample from grain; samples from grain in bags should be taken after thorough mixing.

It is possible to sample the grain flowing from the spout of a combine harvester but this method has its dangers — you will probably not sample the first few seconds, or the last few seconds, of the flow from one plot and does large grain tend to come down on one side, small on the other?

Sugar beet roots can be sampled by taking every 10th or 20th root from each row lifted (the sample fraction being calculated to give the required size of sample).

Crops can of course be sampled for such purposes just before harvest, but there are obvious dangers — the moisture content of wheat grain in the field may change by 10 per cent as the dew dries, or harvest may be unexpectedly delayed by weather or the failure of a machine. So, on the whole, unless the sampling includes estimates of yield, it is usually better not to sample for dry matter percentage and such like before harvest.

## 5.7  RECORDING THE WEIGHTS

Make duplicate records of all weighings by use of carbon paper or carbon-backed paper, or some of the specially-prepared papers used for receipts, invoices, etc. (Pushing carbon paper between a pair of sheets for every plot is not easy in wind or rain.) Put the two copies in separate buildings as soon as possible. Avoid the copying of figures except in desperation; if you *have* to copy get someone else to call over with you as a check. The analysis of the results should be done by someone working from one of the original copies of the weighings. A messy but accurate original, mud, tear-stains and all, is infinitely better than a 'fair' copy containing one gross error of copying.

If you find you have written down a wrong figure do *not* correct it by altering the figure — unless the new figure is at least four times as heavily written as the wrong one. It is much better to cross out the wrong figure and write the correct one just above; or, if there is not enough room above, put it at a distance and show by an arrow where it belongs. Correctness, clarity and tidiness are all greatly to be desired but the greatest of these is correctness (and the least is tidiness).

## 5.8  RECORDING THE ORDER OF HARVESTING PLOTS

Combine harvesters working on plots occasionally seem to hold back part of the grain harvested from one plot, keeping it somewhere in their innards, and then deliver it with the grain of the next plot. The first yield is underestimated, the second overestimated by the same amount. Inspection of the plot residuals sometimes suggests this may have happened without anyone noticing at the time. It is useful in this connection to have a record of the order in which the plots were harvested; this can conveniently be made on a copy of the plan while harvest is going on. This also serves to guard against another bad habit of combine harvesters which was reported recently — a tendency to give different estimates of the same standing crop when working in opposite directions; this may have been caused by a slope in the ground or perhaps by the crop's being lodged all in one direction.

# Chapter Six

## SAMPLING

### 6.1 SAMPLING TO ESTIMATE DIFFERENCES CONTRASTED WITH SAMPLING FOR ABSOLUTE VALUES

Most of this chapter will be devoted to the sampling of the plots of an experiment, the main object being comparisons between plots. Sampling of this sort may be done at all stages, e.g. we may require soil pH values for individual plots before any treatments are applied or any crop sown, we may sample the growing crop for chemical analysis or assessment of disease, we may sample the ripe crop for estimates of yield, and we may sample the soil a month or a year after harvesting the experimental crop to assess residual effects of (say) nematicide treatments. In a few cases the object of sampling may be to determine the mean value of some characteristic for the whole site regardless of plots; an example is the determination of soil pH of a site to decide how much (basal) lime is needed before the experiment is laid down. But in my opinion *any* sampling done after treatments have been applied should if possible be done plot by plot so that comparisons between plots and treatments can be made later if necessary.

In the sampling of an experiment where precise comparisons between plots or between groups of plots are the object, rather than a precise estimate of the mean value for the whole experiment, we are led to methods of sampling different in important respects from the methods appropriate to surveys where the main object is to answer questions like 'what is the mean yield of wheat in England?'. Methods of sampling for surveys will not be discussed in this book. Nevertheless you may apply most of the ideas in this chapter to 'site-sampling' if you regard the site as a single 'plot' for this purpose. If an experiment is sampled by an appropriate method the results will be comparable with those obtained on other experiments sampled by the same method; comparisons between experiments will be as precise as possible for a given amount of sampling. But in the whole chapter my main object is to help you to get accurate estimates of *differences*, not absolute values.

To give an example: if method A has a constant bias believed to be about 10 per cent and a standard error per sample of 4 per cent, while method B has no bias but gives a standard error of 8 per cent you should choose method B for a survey, but method A for an experiment.

## 6.2  OBJECTS OF SAMPLING

The objects of sampling a field experiment are:

(1)  to estimate differences between treatments (in whatever attribute is being measured on the sampled material). We want the most accurate estimates we can get for a given amount of work;

(2)  to provide estimates of error (standard errors or something similar) for the estimates of treatment differences;

(3)  to provide an estimate of 'sampling error' (that is, the variability between replicate samples from one plot).

If the experiment is well designed objects (1) and (2) are achieved (or not achieved) together. This is because tests of significance of treatment differences are based on the mean square for between-plot error (calculated from plot-totals or plot-means) *not* on the mean square for between-samples (within plots) error. So, for estimating treatment-differences and testing their significance we need only use plot means (or totals) and work out an analysis of variance with the sum of squares partitioned exactly as in the analysis of yields. (I assume every plot has been sampled — this is not necessarily true; it is better to sample one block than none, and in some cases it may be appropriate to sample some treatments but not all — but see Section 6.6 for objections to this.)

So why bother to estimate sampling error at all? Why, indeed, unless you are comparing your sampling method with another one, actual or proposed. If you want to know, for example, how much more information would have been obtained with double the number of samples per plot, then you need an estimate of sampling error. But very often you don't. The number of samples taken from each plot, like the size of the plot itself, is seldom determined on the grounds of the precision required in the results, often by practical considerations of time and resources. It's worth remembering too that many forms of sampling appreciably disturb the crop on the remainder of the plot. Some plants may benefit because neighbouring plants have been removed but walking on plots and digging up soil usually damages growing plants.

## 6.3 PRINCIPLES OF SAMPLING

Put briefly but rather imprecisely, the two great principles are:

(1) samples must be representative;
(2) samples must be chosen objectively.

(This is the logical order but I prefer to deal with the two principles in the opposite order.)

### (1) Samples must be chosen objectively

This means that in deciding the exact place at which you are going to take a sample you must be in *no* way influenced by the appearance of the crop or the soil at that place; in the extreme you must accept a sample area that has no crop-plants growing on it and record it as such. The usual way of choosing sampling points is to settle in advance two co-ordinates — that is, the distance along and across the plot from one chosen corner. These can be in feet, links, metres, etc., or in paces (but if the latter do your pacing with your head up, not looking at the crop!). If the crop is sown in rows, so many per plot, it is usual to use the number of rows as a measure in one direction, partly for convenience, partly for reasons given in the next paragraph.

### (2) Samples must be representative

Conditions vary from point to point within a plot and as far as possible samples should be chosen so that the whole range of conditions is represented. Some types of variation are known (e.g. edge-rows are different from inner rows; and rows that have been run over by the wheels of the spraying tractor are different from those that haven't). Some may not be present but probably are (e.g. in most multi-row drills different spouts usually sow seed at different spacings or depths). And some may be there if we are unlucky (e.g. the north end of all plots of a block may be on land of better fertility that the south end).

Finally there may be (in fact, there always is) a variation from point to point over short distances, more or less at random, within each plot — small-scale patchiness.

Let us for the moment ignore the last-mentioned type. With any combination of the other types of variation comparisons between plots will be most precise if the pattern of sampling is exactly the same on all plots, that is, if we use the same sets of co-ordinates (one pair for each sample) for each plot. Any deviation from the fixed pattern will tend to increase the between-plots variance. Now small-scale patchiness affects different plots differently and will increase the between-plots variance whatever pattern of sampling is used. The use of a fixed pattern will (on average) neither increase nor decrease the effect of small-scale patchiness on variance.

So I conclude that, unless sampling errors *must* be estimated, fixed-pattern sampling is best. It is certainly easier — but I am not merely trying to justify my laziness!

The pattern chosen should give a reasonable spread over the length and breadth of the plot and should take into account possible variations between drill-rows and also such things as wheelmarks. Of course you may choose to avoid wheelmarks in deciding your sampling pattern but if areas affected by wheelmarks are included at harvest the relevance of your sample data to yields (e.g. for analysis of covariance) will be lessened. *Figure 6.1* shows some examples.

If you want to estimate sampling error some sort of compromise may be possible, e.g. divide the plot into quarters and take two samples per quarter-plot, located at random; or take two samples located at random within wheelmarked areas, and two at random from unaffected areas. The former scheme will give four degrees of freedom per plot for sam-

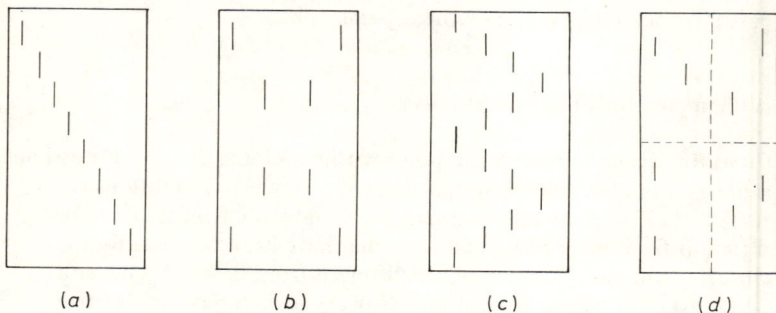

*Figure 6.1* Some methods of sampling plots. (a), (b), (c) are typical schemes of 'fixed-pattern' sampling. (d) is random sampling within quarter-plots ('stratified random sample')

pling error, the latter two. In an experiment with 30 or more plots either scheme will give a good estimate of the sampling error associated with that scheme.


## 6.4  SAMPLES FROM YIELD AREAS OR DISCARDS?

If there are large enough discard areas in your plots, that is, areas that have received the correct plot treatment but are not required for the final estimate of yield, you may be able to take all the samples you want from the discards. (I am thinking chiefly now of sampling plants or soil during the growing season.) If you sample the discards you obviously lessen the correlation between the conditions of the small sample area and the larger harvested area. But the sample area and the harvested area are different, whether the former lies within the boundary of the latter, or outside it; once you've sampled a particular square foot destructively, that square foot cannot contribute to the harvested yield. If you are interested in the effects of the treatments on the variate that is estimated by sampling (e.g. percentage of wheat plants with severe take-all) the best thing is to spread the samples equally densely over the harvested and discard areas — for this purpose the sampled areas need to represent the whole treated area. If, on the other hand, your main concern is the relation between yield and the sampled variate (e.g. if you want to know the average loss of yield of wheat for every additional plant per cent that has take-all) you have to choose between two evils.

(1) Take all samples from the yield area. This gives the maximum correlation between the mean conditions in the sampled areas and the mean conditions in the yield areas. But your yield areas are lessened by the amount of sampling and so their intrinsic variability is increased — the increase, of course, may be very slight. But there are other factors working to increase the variability of the yields and it may be difficult to put even an upper limit on their possible effects. First, you may trample or otherwise disturb many plants in removing your samples. Secondly, if you sample some time before harvest, plants near those removed will probably respond to their extra elbow-room by growing extra vigorously. These, and perhaps other factors, will not necessarily affect all plots equally; apart from random variation, they may 'interact' with treatments, e.g. you will probably do more damage walking through a well-grown, dense crop than through a weak, sparse one.

(2) Take all samples from the discard areas. If you do this your yields are (presumably) quite unaffected by the sampling but the

differences between mean sample conditions and mean 'yield' conditions are subject to greater variability. The variance that matters is increased in the ratio

$$\frac{w + b}{w}$$

where   $w$ = variance of mean of samples (within yield area);
$b$ = variance of difference of sample means on yield area and discard area of the same plot ('between').

This is analogous (except that different sized areas may be involved) to 'sampling variance' and 'between-plot variance' in an experiment in blocks of two plots each. You may have data that will give you some idea of this ratio. The regression of yield on the sampled variate calculated from the 'error' line of the analysis will (on average) be less than the true value in the same ratio that the 'error' (i.e. between-plot error) variance of the sampled variate is increased. If the regression is calculated in some other way, e.g. from the 'treatment' line, in which the relevant variance of the sampled variate is greater, the bias of the regression will be proportionately less. In the sort of situation I have in mind (the only useful one, I think) the sampled variate has much greater variance for 'treatments' than for (between-plot) 'error' and so the effect is slight. If the sampled variate does *not* vary significantly between treatments it isn't likely to be much use except perhaps to 'adjust' the yields and so lessen their variance — but this is a long way from the situation we had in mind at the beginning of this section.

## 6.5   ADJUSTMENT OF HARVESTED AREA TO ALLOW FOR SAMPLING OF THE CROP

Bear in mind also that sampling from yield areas, unless it's done carefully just before harvest, introduces an arbitrary element into the conversion of plot yields to yields per unit area. There is the worse risk that plants near to sampled areas will compensate to different degrees according to treatment. To take an extreme example: if you space potatoes at 36 in. x 36 in. neighbouring plants will scarcely be affected by the removal of a sample plant at mid-season but if the spacing is 18 in. x 18 in. there will be substantial compensation. In such a case no single formula for adjusting the harvested area to allow for sampling will be correct and estimates of the effects of spacing treatments will be biased. If the same rule is applied to all plots (and it would be a bold bad man who would dare to specify different rules for a variety of treatments) the yields of the closer spacing will be overestimated relative to those of the wide spacing.

## 6.6  PARTIAL SAMPLING OF AN EXPERIMENT

As we do not know exactly what effect sampling will have on the yield of a plot, it is much better to sample all plots of an experiment similarly. If some blocks are sampled, others not, since the effect of sampling is likely to be correlated (negatively or positively) with yield, the error mean square (i.e. the blocks x treatments interaction) is likely to be inflated; if on the other hand some treatments are sampled, others not, we may adjust yields to allow for the area sampled but this introduces an arbitrary element into the treatment comparisons.

# Chapter Seven

# LONG-TERM EXPERIMENTS

## 7.1 STRENGTH AND WEAKNESS OF LONG-TERM EXPERIMENTS

Long-term experiments are needed to study the long-term effects of agricultural practice, such as *rotation* of crops and use of organic manures and fertilisers.

It may take several years to establish the differences you wish to study — perhaps the effects of leys of different types, or the effect of a massive application of superphosphate intimately mixed with the soil and chemically assimilated by it. Similarly the effects may persist for several years: for example farmyard manure applied to potatoes may give an important increase in yield of the following wheat crop. Long-term experiments (I am thinking now of experiments planned to last say six years or more) bring a new lot of problems before us, some of which are discussed in the remainder of this chapter.

First, since long-term experiments are relatively expensive to run, they are usually much numerous less than short-term experiments. For this reason, and because of the correlation between the yields of successive crops on the same plots (see Section 8.2), a good, even site likely to give great precision is most important. In a series of annual experiments, perhaps in an effort to have all relevant conditions represented, we may accept some sites where we have reason to expect rather greater variability, but such a site would probably be rejected for a long-term experiment.

Long-term experiments need to be in the full control of the experimenter for the whole of their lives, so most are sited on experimental farms, very few on commercial farms. Most long-term experiments are (or should be) examined and sampled regularly for assessment of some or all of the following:

diseases (airborne and soil-borne; the latter are more likely to vary between treatments);
nutrients in soils and crops;
organic matter in soil;
weeds;
soil structure.

Most long-term experiments seem to be 'fundamental' rather than 'applied'. An annual experiment can help a farmer to decide what variety of seed barley to buy for next season; the results of a ley-fertility experiment may well influence future farming in many important but indirect ways, through a better understanding of the effects of repeated generous dressings of fertiliser on yields and of the effects of crop-rotations on soil-borne diseases. It is appropriate then that long-term experiments (being relatively few) should be sampled and studied in great detail, as the results so collected give a basis for generalising the results to a wider range of conditions of soil and climate. Annual experiments, on the other hand, may often be done for the yields only, with only the most cursory inspection for diseases, etc., unless these are the main object of the experiments. If there are enough sites appropriately chosen the summary of yields may well be of sufficient generality.

We next consider some of the types of long-term experiments that can be distinguished.

## 7.2 MONOCULTURE EXPERIMENTS

The most famous (and successful) experiments involving the prolonged growing of a single crop are the 'classical experiments' laid down at Rothamsted by Lawes and Gilbert between 1843 and 1856. Those on wheat, barley and meadow hay have continued till the present. The one on potatoes failed but has given invaluable information on phosphate residues. The one on beans and the one on leguminous herbage crops are the only ones that have vanished.

Ironically, the monoculture experiments on wheat, barley, hay and root-crops all outlasted the one classical experiment on crops grown in rotation. This, after 70-odd successful years, was spoiled by a soil-borne disease ('clubroot') that affected the swedes. No resistant variety was available, and no suitable substitute crop could be found and eventually the experiment in its original form was abandoned. The site was, however, used for a study of the residual effects of the treatments applied for so long. A long-term experiment, like a chain, is no stronger than its weakest link.

The main advantage of the monoculture experiments is that after the first few years, when soil-borne diseases may increase rapidly (and, in some cases, subsequently decrease and become fairly steady at an intermediate intensity) conditions on the plot or plots of any one treatment tend to be very stable, apart from the slow changes whose study may well be the main object of the experiment. Examples are the progressive enrichment of the soil on plots given certain manures or fertilisers, the gradual depletion of nutrients on others.

Well-managed monoculture experiments provide excellent data for the estimation of seasonal effects on yields of crops. The results of the Rothamsted classical experiments were used by Fisher[27] and others for this purpose. The statistical methods used were elegant and powerful and significant correlations between yield and various functions of the meteorological observations were detected; some of these differed appreciably between plots of different treatments. Yet the value of these analyses to agriculture and agricultural research is slight; certainly much less than their value to statistics. Long-term rotation experiments (to be discussed below) also provide good data for the study of seasonal effects on yield but the results are usually disappointing.

The main disadvantage of monoculture experiments is that ordinary farmers in most areas do not practise monoculture and so conditions in the soil of the experiment may be different in important ways from those of farmers' fields. For example, in the soils of Broadbalk at Rothamsted after 130 years of wheat growing there seems to be an equilibrium between the take-all fungus and some unidentified antagonist so that the damage done to the crop by take-all is less than is often seen on other fields nearby where only three or four successive crops of wheat have been grown.

## 7.3 ROTATION EXPERIMENTS

These involve one or more rotations of crops and are designed so that they can be continued indefinitely.

It is convenient to distinguish two types:

(1) those in which different rotations are to be compared (e.g. tests of ley systems *v.* arable rotations, tests of green manuring);
(2) those in which other factors are tested on one single rotation of crops (e.g. tests of short- and long-term effects of P-fertilisers).

In experiments of type (1) the different rotations have to be accommodated on more or less small plots, preferably with adequate replication; experiments of type (2) usually have all the plots of one crop assembled in one *series*. In both types it is very desirable to have all *phases* of the rotation or rotations present each year. This implies that, once the experiment has completed any necessary preliminary years, a full set of results are obtained each year. If only half the phases are present it will take twice as long to get an equal sample of seasonal effects — and this can be frustrating. Most experiments (like new aeroplanes) are obsolete before they have given much service and you cannot afford fruitless years.

The longer the rotation or rotations of crops involved in one experiment, therefore, the bigger the experiment has to be. This means, for example, that experiments testing the effects of leys usually involve only leys of short duration. (But something can be done by having plots that stay in 'ley' for one complete cycle plus the normal 'treatment' period — as in the ley-fertility experiments of the National Agricultural Advisory Service of England and Wales (see *Figure 7.1* and Horne[34]).

Experiments comparing rotations of different lengths have been tacitly excluded from this discussion so far. They present special difficulties because if the plots are grouped together into blocks there

| Treatment | | 1 | 2 | 3 | 4 | 5 |
|---|---|---|---|---|---|---|
| Year | 1 | H | L | Lu | Lc | L |
| | 2 | P | L | Lu | Lc | L |
| | 3 | O | L | Lu | Lc | L |
| | 4 | P | P | P | P | L |
| | 5 | W | W | W | W | L |
| | 6 | B | B | B | B | L |
| | 7 | H | L | Lu | Lc | L |
| | 8 | P | L | Lu | Lc | L |
| | 9 | O | L | Lu | Lc | L |
| | 10 | P | P | P | P | P |
| | 11 | W | W | W | W | W |
| | 12 | B | B | B | B | B |

*Figure 7.1*  NAAS Ley Fertility Experiment Cropping scheme (L = ley)

will be some years when some of the comparisons one desires to make between the yields of some particular crop involve inter-block differences. In many such cases it may be best to put all phases together in one randomised block (possibly with replicate blocks nearby).

In all other cases it is best to put all plots that are in the same phase together in one block, partly for convenience, partly for increased precision. Replicate blocks in the same phase may be close together — especially convenient in experiments of type (2).

## 7.4  CROP-SEQUENCE EXPERIMENTS

These are experiments planned to last for a definite term of years, usually not more than about eight, sometimes as few as two years. Like rotation experiments they may involve different sequences of crops (e.g. a comparison of different 'break-crops' in a sequence of cereal crops) or one sequence only (e.g. a test of nitrogen fertiliser applied to potatoes, with assessment of residual effects in wheat in the next season). They may include replicates started in successive seasons either in the same field or in different fields.


## 7.5  PLANNING A LONG-TERM EXPERIMENT

I assume that you have decided on the treatments to be compared, the cropping rotation(s) or sequence(s) to be followed, and the arrangement of blocks and phases.

Some special considerations arise in the choice of size and shape of plots. First, all long-term experiments involve treatments that produce differences in soil conditions between plots (if this is not true of your proposed long-term experiment, you should probably be planning a series of independent annual experiments!). If soil moves from one plot on to a neighbouring plot it lessens the value of the results; if the areas taken for yield are seriously contaminated the experiment is probably worthless. The size and shape of plots must be such as to minimise this contamination. Remember also that diseases likely to differ between plots may spread above ground (e.g. stem eelworm or lucerne may be spread in bits of herbage carried along by machinery). The movement of soil can be kept to a minimum by:

(1)  ploughing so as to turn the furrow in alternate directions on successive occasions;
(2)  doing light cultivations along the length of the plots;
(3)  making sure that all implements are sensibly used — for example a rotary cultivator may move soil backwards much or little according to the way it is used.

Secondly, some juggling may be needed to find a plot width that suits all the crops involved (see Section 2.7).

Very often a long-term experiment raises new questions while it is in progress; someone has an idea how the differences in yield already recorded may be explained in terms of different supplies of available nutrients, or moisture, or different intensities of soil-borne diseases and he wants to test his idea. So it is most desirable (I would say essential) to plan from the beginning for the inclusion of extra factors during the life of the experiment. There are two main ways of doing this:

(1) Split the existing plots into sub-plots (e.g. half-plots or quarter-plots) and make the appropriate test(s) on the sub-plots. This is fine for testing something with ephemeral effects (e.g. soluble N-fertiliser or a fungicide applied to the foliage) but (unless the original plots were very large) less satisfactory for testing something like farmyard manure that is likely to have residual effects — the movement of soil will cause contamination between the sub-plots. In some cases P and K fertilisers have been tested on sub-plots for the most important crop of the rotation and equal dressings have been applied to opposite sub-plots after harvest to leave the soil (at any rate in the long run) more or less uniform (see *balancing manuring*).

(2) 'Confound in' extra factors. For example an experiment with two factors at three levels each (3 x 3) arranged in three (or six, nine, . . .) blocks of nine can easily be turned into a 3 x 3 x 3 experiment with the three-factor interaction partially confounded with blocks. $3^n$ and $2^n$ designs lend themselves to this device fairly readily; mixed designs ($3^m$ x $2^n$) do not. In this context a factor at four levels is usually equivalent to two factors each at two levels.

In general, the simpler the original design, the easier it is to incorporate the later tests — which in my experience are often the richest part of the experiment.

Often the experimenter finds he needs to investigate the shape and position of a response-curve — typically the curve of response to fertiliser-N. Four levels seem to be the least that can indicate how much N is needed for maximum yield and what that yield is. Four levels of N can be put into an existing experiment by either of the methods given above, or by a hybrid: split the plots into halves and apply $N_0$, $N_2$ (say) to the halves of some plots, $N_1$, $N_3$ to the halves of the remainder. This may be arranged so as to confound $(N_0 - N_1 + N_2 - N_3)X$ where $X$ is some interaction of the original factors.

The four levels of N may vary according to the original treatment. For instance in a ley-arable type experiment we may test $N_0$, $N_1$, $N_2$, $N_3$ on the *test-crop* following a grazed ley but $N_1$, $N_2$, $N_3$, $N_4$ after an arable sequence where less N is likely to be available from the soil. There are several variants on this, for example,

(1) $N_0$, $N_1$, $N_2$, $N_3$ and $N_0$, $N_2$, $N_3$, $N_4$ (if $N_0$ plots are of special value).

(2) $N_0$, $N_1$, $N_2$, $N_3$ and $N_0$, $N_x$, $N_{2x}$, $N_{3x}$ ($N_x$ might be equal to 1.2 or 1.3 times the unit dressing $N_1$).

In some long-term experiments it may be useful to put in spare plots that can be run on in 'normal' treatment until required; an example is the Rothamsted Cultivation–Weedkiller Rotation (Moffatt[41]). The treatments included when the experiment started were three systems of primary cultivation, ploughing (P) v. rotary cultivation (R) v. tine cultivation (T) in all combinations with three methods of controlling weeds after planting two of the four crops in the rotation (M, X, Y). Spare plots were included; each block contained the nine treatments listed plus three plots which were treated exactly as the PM plots of the main set. After a few years one new treatment (suggested by observations made in an exceptional season) was brought in on one piot per block. Later with the advent of the weedkiller paraquat, another plot in each block was allocated to a 'minimum-cultivation' system that would have been impracticable when the experiment started.

## 7.6 HUSBANDRY OF LONG-TERM EXPERIMENTS

In most crop-sequence experiments and perhaps a few short-lived rotation or monoculture experiments, the experimenter may be content to keep unchanged the husbandry (i.e. varieties, cultivations, etc.) used at the beginning. But in most monoculture and rotation experiments husbandry must be carefully considered beforehand, and reconsidered at regular intervals.

Lawes and Gilbert ploughed Broadbalk with oxen in 1843 and harvested the first crop with sickles in 1844. When oxen gave way to horses and horses to tractors, soil conditions no doubt changed in consequence and the changes may have affected all treatments equally. But more probably the effects of the changes varied, for example between the plot that received farmyard manure annually and the many plots that receive none. When sickles were discarded for the binder and when the binder in turn was scrapped at the advent of the combine harvester, average dates of cutting changed (among other things) and we have no reason to suppose the changes had equal effects on all the plots.

If stability of conditions had been the main object Rothamsted would still be yoking oxen to the plough and sharpening sickles on a stone. Clearly there has to be a compromise between (on the one hand) keeping husbandry constant even when farmers' practices have changed completely and, on the other hand, changing cultivations, varieties, etc., every time a promising innovation is heard of.

A working compromise seems to be to maintain husbandry fixed for one or more periods of, say five years. Towards the end of each period all aspects of practice are reviewed and decisions taken for the next period. This has the advantage that people wishing to plan their observations or sampling have assurance of stable conditions for at least five years at a time. In a rotation experiment it will often be convenient to make changes of husbandry at the beginning of a *cycle*, e.g. in a ley-arable experiment basal or *standard* manuring may be changed as each group of plots are sown with the first *treatment-crops*. This may imply that some series will receive the old manuring while others receive the new; in other words the change is 'phased in'.

As with stocks and shares, it is risky to aim for the maximum gain in the short run; a new variety of wheat that in variety trials has yielded 5 or 10 per cent better than the present one may, after one year of widespread sowing, be devastated by a new race of yellow rust.

## 7.7  AGRICULTURAL JUDGMENT

First, find a good man to give the orders — what implement to use to prepare a seedbed and when to use it. You cannot afford to have one avoidable crop failure in a long-term experiment.

Second, having found him, let him go on making the decisions as long as possible. He will gain experience and his judgment will improve during the course of the experiment (or, of course, he may reach the top of his personal response-curve and start to slide into senility!). But at least you will avoid gross, sudden changes of judgment which would have unknown effects on the yields. The yields of a good long-term experiment form a time-series and their correlation with the weather of the various years may be of value. Also, any slow changes in yield that may be detectable through the seasonal fluctuations may be worthy of study (Dyke[24]).

## 7.8  RESPONSIBILITY FOR A LONG-TERM EXPERIMENT

If the initiative for a long-term experiment comes wholly or mainly from one person, that is fine (but several very good rotation experiments have been hatched and brooded by committees with three or four people taking active part in the discussion). When there is only one really keen sponsor it is worth while for those who are going to have the job of running the experiment for years to come to seek some assurance of continuity; one or more colleagues of the sponsor should be persuaded (or coerced) into accepting a degree of responsibility and taking an

**Fertiliser Plots**

| | Whole | Half | Quarter | Eighth |
|---|---|---|---|---|
| 1961 Red Beet | (N1 v. N2)(P1K1 v. P1K2)(PL1 v. PL2) | (0 v. Mg) | – | – |
| 1962 E. Potatoes/Leeks | (N1 v. N2)(P1K1 v. P1K2)(PL1 v. PL2) | (0 v. Mg) | – | – |
| 1963 Red Beet (1) | (N1 v. N2)(P1K1 v. P1K2)(PL1 v. PL2) | – | – | – |
| 1964 Carrots/Leeks | (N1 v. N2)(P1K1 v. P1K2)(PL1 v. PL2) (2) | – | – | – |
| 1965 Red Beet | (P1K1 v. P2K2) | (0 v. Pt) (3) | (N‴) (3) | – |
| 1966 Carrots | (P1K1 v. P2K2) | (N‴)(N″) | (0 v. SP1)(P1K1 plots) (0 v. SP2)(P2K2 plots) | (S3 v. S4) |
| 1967 Sugar Beet | (P1K1 v. P2K2) | (0 v. Pt) | (N′)(N″) | (N1 v. N2) (4) |

**FYM plots**

| | Whole | Half | Quarter | Eighth |
|---|---|---|---|---|
| 1961 Red Beet | (0 v. N1P1K1) | (0 v. Mg) | – | – |
| 1962 E. Potatoes/Leeks | (0 v. N1P1K1) | (0 v. Mg) | – | – |
| 1963 Red Beet (1) | (0 v. N1P1K1) | – | – | – |
| 1964 Carrots/Leeks | (0 v. N1P1K1) (2) | – | – | – |
| 1965 Red Beet | (0 v. P1K1) | (0 v. D1)(D1 plots) (0 v. D2)(D2 plots) (N‴) (5) | (N′)(N″) | (S3 v. S4) |
| 1966 Carrots | (0 v. P1K1) | (0 v. D1)(D1 plots) (0 v. D2)(D2 plots) | (0 v. SP1) | (N0 v. N1) (4) |
| 1967 Sugar Beet | (0 v. P1K1) | (0 v. D1)(D1 plots) (0 v. D2)(D2 plots) (N‴) | (N′)(N″) | – |

(1) Also (0 v. SD) to columns of 4 half plots.
(2) N rates changed, see under symbols 1964.
(3) Allocated at random (subject to confounding) ignoring (0 v. Mg) 1961 and 1962.
(4) Allocated at random ignoring (S3 v. S4) 1965.
(5) Cumulative FYM (10 tons on D1 plots, 20 tons on D2 plots) v. Residual FYM (no fresh applications) allocated at random (subject to confounding) ignoring (0 v. Mg) 1961 and 1962.

*Figure 7.2* Woburn Market Garden Experiment, treatments and cropping 1961–67 (the explanation of the symbols has been omitted)

active interest from the beginning. They may be working in the same discipline as the sponsor — or in a different one that is involved in the proposed experiment. (It is hard to imagine a long-term experiment that has not something useful to offer to people of at least two different interests.) A long-term experiment whose sole sponsor has left, died, or lost interest is a sad orphan and adoption is seldom quite successful.

## 7.9  RECORDING THE DESIGN AND TREATMENTS OF LONG-TERM EXPERIMENTS (ESPECIALLY ROTATION EXPERIMENTS)

A detailed *plan* will of course be drawn each year and this, together with a full explanation of all symbols used to denote the treatments, tells the experimenter (and others) all he needs to know in any one year. As these records accumulate they contain full information on the treatments and design, but not in a very convenient form. I suggest two additional forms of record can be useful.

(1) A life-history in symbols of every plot, including treatments applied to sub-plots; this is almost essential when new sub-plot treatments are 'confounded in' with older sub-plot treatments. This type of record may require one page per whole plot and so is fairly bulky. It conveniently allows the recording of any errors of treatment of individual plots.

(2) A more compact summary of the design, based on the idea of a design-key (Patterson[43]). My version is a table in which each row represents a year (or run of years in which the design and treatments did not change) and each column represents one contrast (or a group of contrasts) between particular sets of plots. A simple example is

| | | | |
|---|---|---|---|
| 1967 | $(A_0$ $v.$ $A_1)$ | $(B_1$ $v.$ $B_2)$ | $(C_0$ $v.$ $C_1)$ | — |
| 1968 | $(A_0$ $v.$ $A_1)$ | $(B_1$ $v.$ $B_2)$ | $(C_0$ $v.$ $C_1)$ | — |
| 1969 | $(A_1$ $v.$ $A_2)$ | — | $(C_0$ $v.$ $C_1)$ | $(D_0$ $v.$ $D_1)$ |

This indicates that the experiment tested the same 3 factors applied at the same rates and on the same plots in 1967 and 1968. In 1969 plots previously given $A_0$ were given $A_1$, and plots previously given $A_1$ were given $A_2$; factor B was omitted, factor C was continued as before and a new factor D was introduced. How this was done (e.g. by 'confounding in', or by splitting plots) can be indicated by a note. The key given implies that the residual effects of factor B can be examined in 1969 (unless a note to the contrary is included). A more complex example, taken from Rothamsted[46] is shown in *Figure 7.2.*

# Chapter Eight

# INTERPRETATION AND PRESENTATION OF RESULTS

## 8.1 ESTIMATION AND SIGNIFICANCE

The object of almost all field experimentation is the estimation of
effects. (This is equally true of most other experimentation.) The testing
of significance is important but secondary. The above sentences,
executed in poker-work, ought to hang permanently over the desk of
every field experimenter.

If you test a fertiliser, a variety, a rotation of crops or a cultivation,
you want to know 'by how much does this treatment increase (or
decrease) yield relative to some appropriate standard treatment?' — that
is, you want an estimate of the *effect* of the treatment. Next, of course,
you want to test the significance of the difference you have recorded.
That is, if a sceptic comes along and says 'the difference you have
recorded is no greater than the sort of difference you would have got
if you had treated all the plots the same' (the 'null hypothesis') do you
nod meekly or do you say 'no, I don't agree'? The simple situation to
remember for comparison is this.

You toss a coin 20 times and record 15 heads, 5 tails. The sceptic
says 'the difference (from 10:10) is no greater than I would expect
with an unbiased coin!'. Do you agree with him, and accept the null
hypothesis (that the coin is unbiased and in the long run will give heads
and tails equally often) or do you reject it? If you decide to reject the
null hypothesis you are then free to set up some other hypothesis; a
reasonable one in the absence of other evidence is that the coin is
biased to an extent that gives a proportion of heads: tails of 3:1 (i.e.

15:5) in the long run. This corresponds with accepting the mean yields of the treatment as estimates of treatment effects in a field experiment.

In the coin-tossing case, assuming an unbiased coin, it is possible to calculate precisely how often a sample of 20 thrown will give the result 15 heads, 5 tails or one of the more extreme results (16:4, 17:3, 18:2, 19:1 or 20:0). The probability that one or other of these results will turn up is about 0.02. In other words, if you record the result of 20 throws many times you will get one of the specified results in about 2 per cent of cases. It is normal to consider results deviating from the 'expected' (10:10) in either direction so we end up with the statement: 'In repeated samples of 20 throws of an unbiased coin a result deviating from expectation as much as or more than the observed sample (15:5) will occur about 4 times in 100 trials'. Most people accept a statement of this sort with a percentage of 5 or less as good evidence that the null hypothesis is disproved. Often such a level of significance is indicated by one asterisk (*). If the percentage is 1 or less two asterisks are used and (occasionally) if it is 0.1 or less three asterisks. I have written above mainly in terms of percentages; the usual convention is to record levels of probability (symbol '$P$'); thus '$P < 0.05$' corresponds to 'less than 5 per cent' of repeated trials.

Bear in mind that, if you are making many tests of significance some are going to give a positive (significant) result *by chance alone*. Put another way, if you do a large number of *uniformity trials* and analyse them pretending that different treatments *had* been applied, then the mean square for 'treatments' will be significantly greater than the mean square for error in a proportion of the analyses. If you are using the 5 per cent level as your criterion of significance, then in the long run 1 analysis in 20 will achieve significance; if you use the 1 per cent level, 1 in 100, and so on.

In the first draft of this book at this point I put in an example — an imaginary experiment whose primary object was the testing of significance. It took me some time to think of it and I had to dive into genetics. A critic better informed than I in the subject pointed out a serious flaw. I have again tried hard to find a suitable example, but in vain. Surely there *must* be a case where tests of significance matter more than estimates of effects?

Now for a possible objection to my dogmatic poker-work statement. The question 'which of these treatments gives the greatest yield?' is a legitimate aim of an experiment. But the answer depends on the treatment means and not on a test of significance. In effect we are asking 'does treatment A give more or less yield than treatment B?' and, if there are many treatments, a lot of similar questions. The tests of significance that can accompany each of these questions of course influence our attitude to the answer; the difference is comparable to the difference between the punter who uses a pin (he has no data to support his choice) and the one who studies 'form' (and believes the data more or less 'significantly' indicate the best horse). Finally, remember there are two types of error in this matter: putting too much confidence in a difference that is large, forgetting that it is not significant; and putting a lot of confidence in a difference that is significant, forgetting that it is too small to be of practical importance. The first can arise when you have an experiment with little replication and large variability, the second when you have an experiment with plenty of replication and small variability. Beware of both. Tests of significance (like electronic computers and the writers of textbooks) are good servants but bad masters.

## 8.2 CORRELATION BETWEEN TESTS OF SIGNIFICANCE BASED ON THE SAME SET OF PLOTS

One final point about tests of significance. You may weigh several types of produce from the same crop (e.g. grain and straw of wheat, roots and tops of sugar beet). Each of the 'variates' so measured may be analysed, giving standard errors and tests of significance. Remember that these variates are almost certainly correlated positively, because plots which by chance have above-average fertility will tend to give above-average yields of grain *and* straw, or of roots and of tops. This means that tests of significance made on the several variates from the same crop are not independent. But we seldom know the numerical value of the correlation. It is likely to be between 0 (when the tests *are* independent and 1 (when the analysis of the second variate is merely a reflection of that of the first scaled up or down by a constant factor). So beware tests of significance on several variates based on the same plots. A safe (but perhaps over-rigorous) rule is to do analysis of variance on one variate only for any one crop.

When in a long-term experiment yields are recorded for the same plots for several years the same point applies. If different crops are grown from year to year the correlation of yields may be smaller than if the same crop is grown repeatedly (though this is not necessarily true). If the plots carry a perennial crop (e.g. apple-trees) the existence of a correlation is very likely – it may be positive if for example the dominant cause of variability is the size of trees but it could I suppose be negative if the trees were of a biennial cropping habit but not all in phase. The same sort of thing can occur in repeated cropping with wheat. Take-all disease increases to a maximum in the first 2–4 years and then declines, with yields behaving inversely; if different plots are out of step a negative correlation between yields in successive years may be generated.

## 8.3 TABLES

The report of a field experiment must include adequate tables of mean
yields. There is usually no need to report the yields of individual plots.
In an experiment with only one factor (e.g. varieties) the mean of each
treatment should be given, and the general mean. You may wish to give
figures or tables derived from the simple table of treatment means. For
example the mean of all tall varieties and the mean of all short varieties,
or, in an experiment comparing different forms of nitrogen fertiliser,
the difference of each treatment from the mean of the untreated plots.

For all experiments (factorial or otherwise) appropriate standard
errors should be given for all figures in the tables, and any derived
figures. The general mean has *no* standard error (because there is no
figure to compare it with − it is correlated with all mean yields in the
tables, and for comparison with means, etc., of other experiments a
standard error based on the standard error (S.E.) per plot is *not* valid).
Also, in some experiments some figures (e.g. means of treatments that
are applied to blocks) needed in the table, cannot be given a standard
error.

In reporting factorial experiments there is more latitude. Clearly,
the means for the levels of each factor averaged over all levels of all the
others must be presented (to exhibit the main effects). Equally clearly
the two-factor interactions must be stated − otherwise there was no
point in doing a factorial experiment. This can be done either by giving
means for all combinations of any two factors (averaged over the rest)
or, in $2^n$ experiments, by giving a table of differences. For example, a
$2 \times 2$ experiment might be reported thus:

|  |  | Factor B | | |
|---|---|---|---|---|
|  |  | *Absent* | *Present* | *Mean* |
| Factor A | *Absent* | 20 | 32 | 26 |
|  | *Present* | 24 | 40 | 32 |
|  | *Mean* | 22 | 36 | 29 |

(Note: it is usual to make rows of the table correspond to the first
    factor, A in this example, and columns to the second, B.)

|  | *Mean* | A | | B | |
|---|---|---|---|---|---|
|  |  | *Absent* | *Present* | *Absent* | *Present* |
| Effect of A | +6 | − | − | +4 | +8 |
| Effect of B | +14 | +12 | +16 | − | − |

Provided that the general mean (29 in this case) is given with the table of differences it is easy to reconstruct the two-way table of means with marginal means given above. (And equally, of course you can calculate the table of differences from the two-way table.) Each form of presentation requires standard errors. If the experiment was straightforward (i.e. had no confounding, missing plots or such like) two S.E.'s are needed in each presentation. For the two-way table we need one S.E. for each of the entries in the body of the table (20, 32, 24, 40) and another for the marginal means (26, 32, 22, 36). If the former is $s$, the latter is $s/\sqrt{2}$. The table of differences needs one S.E. for the mean effects (synonymous with main effects) and another for entries in the remaining columns. These are respectively $s$ and $s \times \sqrt{2}$. Either form of presentation can be easily extended for experiments with three or more factors each at two levels ('$2^3$', '$2^4$', etc.). Each omits information on three-factor interactions, four-factor interactions and so on. If you wish to give this information you have two choices:

(1) Give (as well as two-factor tables) all three-factor tables, four-factor tables, etc., ending up with a table of the mean yields of all the treatment combinations included in the experiment (but 'ware confounding — see below).

(2) Give the mean and a full list of main effects and interactions, e.g.

Mean = 84
A     = +10 (i.e. main effect of factor A)
B     = +8
C     = −7
AB    = −2
AC    = +3
BC    = 0
ABC   = +1

Confounding can be dealt with more easily in this presentation, but it assumes a fair degree of sophistication in your readers. They (and you) should be familiar with the manipulations of effects and interactions — see for example Yates[8].

Experiments with one or more factors at three or more levels are usually best reported by tables of means (two-factor tables with marginal means, perhaps also three-factor tables; you *can* go on to four-factor tables but these may dismay two readers for every one who uses the extra information of the very complex tables). Experiments such as $3^2$ or $3^3$ can be reported in terms of 'linear' and 'quadratic' components and their interactions but here again we are approaching a stratosphere where only the well-acclimatised reader will breathe freely.

## 8.4  CONFOUNDED EXPERIMENTS

If your experiment had any interaction(s) totally confounded you need
a convention. (Partial confounding is another matter which I shall not
deal with — see Yates[8].) The convention usually observed is to assume
each confounded interaction is zero and to adjust the yields so that the
block means are equalised, by adding to or subtracting from each plot
in a block the same amount. For example, if an experiment gave the
following plot-yields (I give one replicate only for simplicity):

| Block | I | | II | |
|---|---|---|---|---|
| Treatment | $a_0b_0$ | $a_1b_1$ | $a_0b_1$ | $a_1b_0$ |
| Yield | 8 | 10 | 9 | 13 |

We should give adjusted yields:

| $a_0b_0$ | $a_1b_1$ | $a_0b_1$ | $a_1b_0$ |
|---|---|---|---|
| 9 | 11 | 8 | 12 |

the block means are 9, 11, difference (II − I) = 2; we therefore add 1 to
each yield in block I, subtract 1 from each yield in block II. Note that
the interaction AB [which in an unconfounded experiment is estimated
by $\frac{1}{2}(a_0b_0 + a_1b_1 - a_0b_1 - a_1b_0)$] has been 'adjusted' to zero.
  A snag is that, in any experiment with confounding, there is no
single standard error applicable to all comparisons between pairs of
means.
  Another slight snag is that some of your readers may think your
block differences are useful estimates of the true interactions (e.g. a
reader who is summarising many experiments may have many such
inter-block comparisons available to average). If you give adjusted yields
only he cannot do this.

## 8.5  UNCONVENTIONAL COMPARISONS

Digressing a little, remember that *any* comparison is better than none.
Inter-block comparisons are better than inter-field comparisons but
given enough inter-field comparisons, a treatment with a large effect
will show a mean difference that may be significant when compared
with the 'treatment x fields' interaction.
  Comparisons between yields obtained in different years on the same

land are available from many long-term experiments and these inter-year comparisons given adequate 'replication' (i.e. enough years with each treatment) and a constant standard of husbandry, can give useful information (Dyke[24]).

## 8.6  HOW MANY DECIMAL PLACES?

Do not put more significant* figures in your tables than strictly necessary. When deciding how many figures in your tables bear in mind:

(1) The relevant standard error; if the standard error of the most accurate figure in the table is ±0.1 there is no point in giving any decimal places in the mean yields [but see (4) below]. (By the way, all entries in a table of yields should be given to the same number of decimal places and all standard errors should be given to one more decimal place than the means; this allows you to round off accurately after multiplying by 't' or $\sqrt{2}$ or some such factor.)

(2) The more superfluous figures appear in a table the greater the risk of a gross error in manuscript, typescript or printed paper.

(3) The more superfluous figures appear in a table the harder it is for a reader to see the important differences — and the less likely a hasty reader is to see them at all.

(4) On the other hand, someone may want to use your data to multiply by some factor (e.g. to convert from '85% dry matter' to 'dry matter' and one extra figure will enable him to avoid rounding errors that might possibly matter. I think the best advice is: if in doubt, leave it out.

If your units (e.g. kilograms per hectare, alias kg/ha, alias kg ha$^{-1}$) are such that your yields are of the order of 1000 or more (with a standard error of say ±20) I suggest you take out a factor of 10 (or 100, etc.) once for all.

*Yield in 10's of kg/ha*

| Treatment A | 99  | |
|-------------|-----|------|
| Treatment B | 112 | ±2.6 |

---

* In the arithmetic sense, not the statistical sense.

is far better than

(*a*)

|  | Yield in kg/ha |  |
|---|---|---|
| Treatment A | 994 | } ±26.3 |
| Treatment B | 1123 | |

or (*b*)

|  | Yield in kg/ha |  |
|---|---|---|
| Treatment A | 990 | } ±26 |
| Treatment B | 1120 | |

(*a*) has one totally superfluous digit in all entries; (*b*) is ambiguous, does 990 mean 'something between 989.5 and 990.5' or 'something between 985 and 995'?

I do not expect much support for my own preference: all tables to be presented in arbitrary units ('arbs') chosen so that as many as possible of the table entries are of the form *ab.c* (that is, from 10.0 to 99.9). The table would end with a definition of the 'arb' in imperial and metric units, e.g.

$$1 \text{ arb} = 0.3 \text{ cwt/acre} = 377 \text{ kg/ha}$$

The reader interested in the main features of the results would ignore (or even score out) all figures after the decimal point, but any one basing calculations on the means given could use the extra figure if he wished.

## 8.7 'WARE PERCENTAGES

Please give *yields* in your tables. If you want to give tables of 'percentage of untreated yield' or 'percentage of control variety' by all means do — but in addition to, *not* instead of the actual mean yields. There are two reasons for this recommendation.

(1) It makes things easier for anyone who wishes to use your results averaged with others.

(2) It helps your reader to avoid spending a lot of time digesting results that he may later find are of little practical application because the yields were much less than those commonly recorded by commercial growers.

## 8.8  STANDARD ERRORS

Your report should include a statement of the standard error per plot
(and the number of degrees of freedom on which it is based) and it is
usual to give this also as a percentage of the mean (i.e. the coefficient
of variation) though if you omit this the reader can calculate it for
himself. In adequately replicated experiments with split plots there
are two or more mean squares for error and you should give standard
errors per whole plot, per half-plot, per quarter-plot or whatever is
appropriate. In such experiments some table entries have two different
standard errors, one for comparison with other sub-plot treatments
within the same whole plot treatment or treatments, the other for
comparison with a treatment-combination not in the same whole plot.

This makes exact tests of significance very complicated (when a
standard error is based on the mean square for error for whole plots
*and* the mean square for error for sub-plots the simple 't' test does not
apply) and if the S.E.s for the two types of comparison are not very
different I give an average figure and call it 'standard error (approximate,
for all comparisons)'. Remember that in most experiments the stated
standard errors have standard errors of at least 10 per cent. (You would
need 50 degrees of freedom for error to achieve this.) I would rather
my reader used an approximate standard error than ignored two exact
ones.


## 8.9  COUNTS, SCORES, CHEMICAL ANALYSES, ETC.

Any set of observations made on all plots of an experiment can be
analysed and tabulated with standard errors in the same way as the
yields. An example showing how visual scores can contribute to our
knowledge of the action of treatments is given in *Figure 4. 7.*


## 8.10  USES OF COVARIANCE

Analysis of covariance may be appropriate in two distinct situations
involving yields and subsidiary observations.

The first (and simpler) case is that in which the subsidiary observations cannot conceivably have been affected by the treatments. For
instance, we might use as subsidiary observation (or 'independent
variate' in the language of statistics) the yield of a preliminary crop
grown on the plots before the treatments were applied, or the amount
of citric-soluble P in samples of soil taken before the treatments were
applied. 'Taking out the regression' of yields ('dependent variate') on

the independent variate (which is all that analysis of covariance usually amounts to in such cases) may result in a smaller mean square for error and so increase the precision of the experiment. Two points are worth noting at this stage.

(1) The fact that the subsidiary observation $x$ is not affected by treatments does not imply that the mean values of $x$ for the various treatments are the same; in general they are not — the $x$'s of any one treatment are, apart from block or row and column differences, a small random sample from a certain population and their mean is subject to sampling error. Covariance analysis, besides giving (we hope) a smaller mean square for error, gives us the ability to adjust each treatment-mean of the yields ($y$) to allow for differences in the treatment-means of $x$.

(2) Covariance may *not* improve the precision of the experiment; if it does not lessen the error mean square it will in fact lessen the precision because of the loss of one or more degrees of freedom for error, giving a less-precise estimate of the population variance. So don't do covariance analysis just to show off!

The second case is completely different logically, though the appropriate analysis is arithmetically rather similar to that just discussed. Now we suppose $x$ is a measurement that is, or is likely to be, or is just conceivably, affected by the treatments of the experiment. Here are some examples:

(1) $x$ = number of live aphids per plant after the application of insecticide treatments;
(2) $x$ = number of sugar beet plants in an experiment including fertiliser treatments applied before sowing;
(3) $x$ = pH of soil measured after applying lime at different rates;
(4) $x$ = yield of crop in an earlier year of a long-term experiment in which some or all of the treatments are applied repeatedly.

In such cases an *appropriate* analysis of covariance will help us to understand to what extent $x$ is a measure of the effect of the treatments; put another way, it may enlighten us on the sequence

$$\text{treatment} \rightarrow x \rightarrow \text{yield}$$

Put crudely it tells us how much of the effect of any one treatment or factor can be explained in terms of $x$. There is a risk here that uncritical use of covariance will 'explain away' *all* the treatment-effects. In the first example just given it is possible that analysis of covariance of yields on aphid counts may leave us with no significant effects of treatments, although if we ignore the counts and do a straightforward analysis of variance we shall find highly significant effects. The mean yields, which before 'adjustment' showed large differences, will show much smaller differences after 'adjustment' by regression. This is equivalent to losing the baby down the plughole with the bathwater.

Another pitfall is the temptation [in an experiment whose treatments can conceivably have affected the number of plants as in example (2) above] to 'adjust' all yields to a constant plant density.

[You are of course too astute to fall into the even deeper pitfall of adjusting for differences in plant density by calculating yield per plant and using that as a basis for analysis; you know as well as I do that widely-spaced plants grow better because they have less competition than closely-spaced plants and that in consequence a loss of $x$ per cent of plants results (usually) in a loss of yield of less than $x$ per cent.]

Now consider a case in which a complex analysis of covariance might lead us to an improved interpretation of the results of an experiment. The experiment has as factors:

lime at 4 rates $(L)$

P at 2 rates $(P)$

K at 2 rates  $(K)$

Soil samples are taken after the lime has been applied and incorporated and pH values are determined. We wish to use analysis of covariance to investigate the value of pH as a measure of the effect of lime on yield. (This is based on a real experiment but for clarity I shall use imaginary results.) Let us use $x$ to denote soil pH, $y$ for yield. We will assume that $x$ varies significantly with the rates of lime (if it doesn't something is very wrong!) but not with $P$ or $K$.

*Figure 8.1* illustrates several types of patterns we may discover in the relation between yield, treatment and pH. In all these diagrams I have ignored block-differences. If these are too big to ignore the $y$'s and $x$'s can both be adjusted for them before plotting, so the diagrams serve equally well as examples. They are all imaginary and I have been at pains to eliminate the sort of overlapping and stray outlying points that turn up uncomfortably often in real life. You may reasonably say 'Why bother with covariance in such cases? Surely we can get all we want by drawing lines by guesswork on a diagram?' To this I reply 'Go ahead, and if all your experiments have such a happy outcome, you are lucky. But what will you do when a doubtful case arises, when the $x$-values of the treatments overlap and there's a *hint* that the various regressions are not exactly the same?'.

In *Figure 8.1a* we have the simplest case (barring the one where there is *no* covariance and $y$ is independent of $x$). Here one single

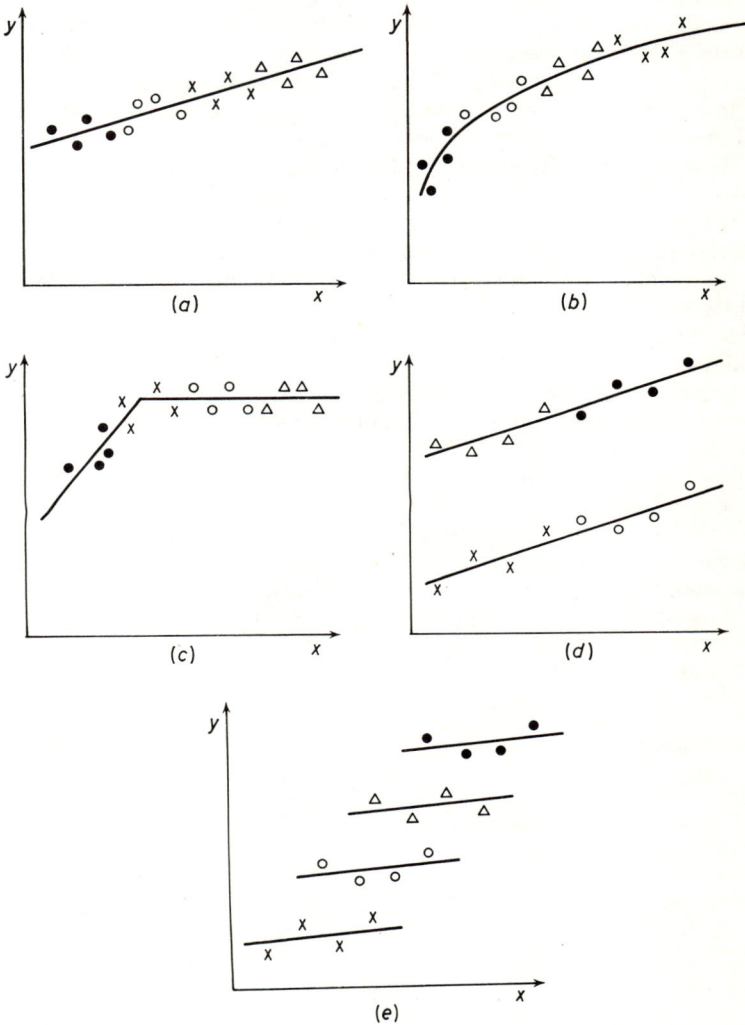

*Figure 8.1* Illustrations of covariance. In each diagram yield (*y*) is plotted against a covariate (*x*) that has been affected by the treatments. Four treatments are included represented as follows: Treatment 1, ●; 2, O; 3, X; 4, △

(linear) regression equation gives a good fit to *all* recorded yields regardless of treatment; $x$ may be said completely to explain the effect of the treatments (and much of the random variation too). We may be able to make a statement such as 'for each unit increase of pH yield has increased by 5 cwt per acre' — a useful practical statement.

In *Figure 8.1b* again a single regression equation gives a good fit but the equation is clearly *not* linear — a quadratic equation involving $x^2$ as well as $x$ will give a better fit (though some other equation may be better still). In *Figure 8.1c* the situation is slightly different in that variation of $y$ with $x$ is apparently linear up to a certain value of $x$; for greater values of $x$, $y$ does not vary with $x$. (This illustrates the doctrine of 'limiting factors' — sometimes very useful.) In this case we may be able to say 'for each unit increase of pH up to pH 7 yield has increased by 5 cwt per acre but increases of pH beyond pH 7 have not affected yield'.

In *Figure 8.1d* the variation of $y$ can be interpreted in terms of two parallel straight regression lines, one applying to one set of treatments, one to the remainder. (This might arise, but with reversed slope, if $x$ were a count of nematodes and the treatments were nematicides some of which also provided a nutrient.) In *Figure 8.1e* I show another case where the regression of $y$ on $x$ *within* treatments is not sufficiently marked to explain the differences between treatments; here there is some point in estimating the 'between-treatments' regression independently. Finally in *Figure 8.2* I show what happened

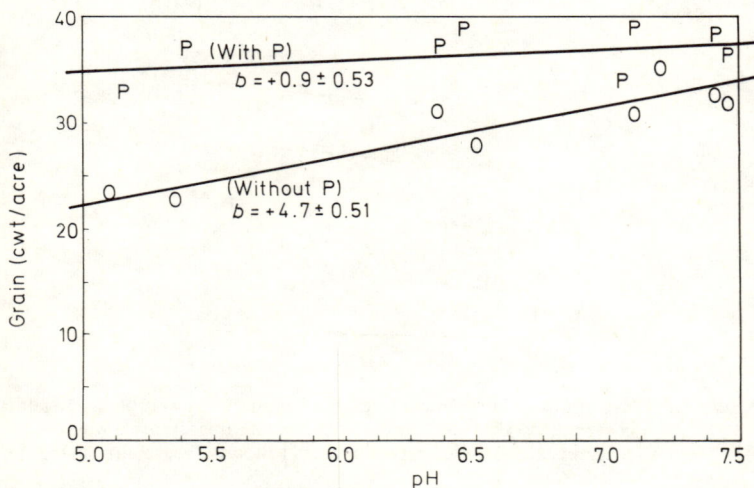

*Figure 8.2* Woburn long-term lining experiment, barley (1967). Regressions of yield on pH, based on treatment-means

in the real experiment on lime, P, K. There are plenty more possibilities quite likely to occur in practice.

For the mathematical basis of analysis of covariance see Chapter 14.


## 8.11  BACKGROUND INFORMATION

Miscellaneous records, some or all of which may be usefully added to your report on an experiment include:

(1)  location of the experiment;
(2)  crops grown on the site for at least two years before the experiment;
(3)  basal dressings, sprays, etc.;
(4)  dates of major cultivations;
(5)  dates of application of treatments;
(6)  dimensions of each plot (or area);
(7)  variety of crop and seed-rate;
(8)  exact treatment of 'untreated' plots (e.g. in a spraying experiment were the untreated plots sprayed with water, water plus wetter or nothing?).


## 8.12  GROUPS OF EXPERIMENTS

If you have the results of a single experiment the standard errors you attach to the mean yields of the various treatments, and the tests of significance you may apply to treatment-differences are a guide to the reliability of your findings; to be more precise, they indicate the degree of confidence with which you may expect similar results if the experiment is repeated *in exactly the same conditions.* The site must be the same, the incidence of pests and disease the same, the weather the same (and so on) as in the experiment you have done. You will agree, I'm sure, that such a repetition of a field experiment has never been achieved and never will be this side of Armageddon. This gloomy statement should be remembered by people thinking of publishing the results of a single experiment; I am not saying that you should never publish the results of one experiment but that you should recognise the risks involved in doing so. If you have the results of several experiments done on different sites in the same year new problems of interpretation arise. Before we discuss them, a minor point needs to be dealt with.

The experiments need not be identical in design or treatments or replication, some may be in randomised blocks, some latin squares;

some may have additional treatments (this often occurs in variety trials, varieties of local interest being added at individual sites). If some experiments have additional factors (as distinct from additional treatments) you will need to be cautious. For example, in a series of experiments testing N, P and K for potatoes, in which some sites had none $v$. farmyard manure as an additional factor (the rest having no FYM) the main effect of K would probably be less on the sites where half the plots received FYM and you would probably be wise to calculate the responses to K in the absence of FYM and use these in summarising K-effects over the whole group of experiments.

By such means we assemble our data for summary. In effect we have, for each site, a mean yield and estimates of the effects and interactions of the factors tested with appropriate standard errors (based on the standard error per plot of each experiment). There are various processes of estimation available: we can, for example, calculate the (unweighted) mean of each effect and interaction over all sites or we can divide up the sites into classes according to soil type or previous cropping, or variety of the experimental crop, or some combination of such factors and calculate means for all classes. We can also divide the sites up according to the mean yield using two or more classes. Alternatively, we may calculate weighted means. Each estimate can be given a weight inversely proportional to its estimated variance [i.e. $1/(SE)^2$ where SE indicates the standard error of the estimate].

At first sight, the latter alternative seems advantageous; it is thrifty with information because it allows extra value to the sites which gave most information and minimises the 'disturbance' due to a few sites which gave especially inaccurate estimates. There are two arguments against the use of weighted means of this sort:

(1) As we have seen earlier standard errors are themselves subject to sampling error and, unless the numbers of degrees of freedom involved are very great standard errors derived from different experiments can vary substantially even though the population standard deviations are equal. Two standard errors based on six degrees of freedom each are as likely as not to be in ratio 4:3 or worse. A test for homogeneity of estimates of variance is available (this simplifies to the F-test when only two estimates are available); see Bartlett[14].

(2) There may be a correlation (positive or negative) between the intrinsic variability of a site and its responsiveness to one of the treatments or factors on test. For example, P-deficient sites might well give greater standard errors than those with plenty of available P in the soil. In this case weighting the results of each experiment in inverse proportion to the estimated variances

of the effects would cause us to underestimate the mean effect of P over all sites.

We are, in fact, between the devil of inefficiency and the deep blue sea of bias; two compromises may help us to placate the devil without being drowned.

(1) Reject all experiments whose standard error per plot is greater than some arbitrary value, or perhaps better, reject all whose coefficient of variation (i.e. SE%) exceeds some arbitrary value. (If the experiments vary in design or replication I suppose you could set a limit for SE (or SE%) per treatment rather than per plot.) Use all acceptable experiments to calculate unweighted means.

(2) Give each experiment a weight inversely proportional to the estimated variance of each treatment but put arbitrary upper and lower limits to the weights used, so that experiments with very small standard errors get less than their calculated weight, those with very large standard errors more.

All this is fine to write about, but a lot of subjectivity is creeping into our summary of the results of a series of experiments. I believe that a cardinal virtue of good planning (whether of one experiment or of a series) is that there should be one and only one interpretation of the results. This brings me to my general recommendation:
Use unweighted means of the results of all available experiments unless you are sure that this is wrong.
Having dealt, not very satisfactorily, with estimation we turn to standard errors and tests of significance. Suppose, for the moment that your experiments were all done in one year, on a variety of sites. There are two courses open to you, each valid in its way, and you have to choose.

(1) You may calculate a mean mean square for error, that is tabulate the error mean squares of all the experiments and average them. (If they are based on different numbers of degrees of freedom it is better to add together the sums of squares and divide by the total number of degrees of freedom.)

(2) You may tabulate each treatment-mean for each experiment and analyse the variance of these, taking out from the total sum of squares the contributions due to 'treatments' and to 'sites' leaving a residual sum of squares which is formally the 'treatment x sites' interaction. It has $(t-1)(s-1)$ degrees of freedom if there are $t$ treatments and $s$ sites. From this mean square ($E$)

you calculate the standard error of each treatment mean in the usual way, the formula being SE (treatment mean) = $\sqrt{(E/s)}$.

A qualification must immediately be added. Beware lest some treatment-comparisons are more variable between sites than others. For example in variety trials disease-susceptible varieties may give yields that vary more between sites than those of disease-resistant ones and so a comparison between a resistant and a susceptible variety may vary more than a comparison within either group. Or, in a factorial experiment, the main effect of one factor may vary more between sites than the effect of another. If there are enough sites you can calculate a separate error mean square for each treatment, comparison, or main effect and interaction (or you may group the comparisons into two or more groups). This is fine until you come to make complex comparisons among the final means; you may then need to allow for correlations between the separately estimated errors — in fact you may end up making a separate calculation for every comparison you wish to examine.

In many cases the two courses of action will produce very different standard errors; most treatment-differences vary far more between sites (let alone between seasons) than between replicate blocks within a site. So the choice is important and it is *yours*. (Your guiding statistician may, like a judge, direct you pretty strongly towards a verdict, but the ultimate responsibility lies with the man who writes the report.)

The basis for the choice is this:

If you are interested in the validity of your results for an imagined repetition on the same sites in the same season, choose (1) — pooling within-experiments errors. If you are interested in the validity of your results for the population of fields of which your sites may be regarded as a sample (still in the particular conditions of the season in which the experiments were done) choose (2) — using 'treatment x sites'.

Finally, if your series involved several sites in each of several seasons (the individual sites in each season may be correlated with those of other seasons, e.g. a succession of different fields on a fixed set of farms, or they may not), you have the chance to estimate 'treatments x sites' and 'treatments x years' and 'treatments x sites x years' mean squares separately. These estimates may perhaps be pooled if they do not differ much or they can be combined in different ways to give standard errors appropriate for applying your mean results to other sites in the same years, other sites in other years. Here you should seek expert guidance, *not* from this book.

One last point about groups of experiments: work with yields per unit area *not* yields expressed as percentages of some chosen treatment. The latter procedure gives undue weight to experiments with small mean yields and too little to those with heavy yields. (If you *must* have this sort of bias it's generally better to have it in the opposite sense in the hope that the farmers who are clever enough to achieve heavy yields will have more leisure to consider and more initiative to apply your results.)

# Chapter Nine

## CRITIQUE OF TECHNIQUE

### 9.1 GENERAL REMARKS

The techniques used at all stages of doing an experiment should always be under critical review in the experimenter's mind. Occasionally it may be worth while to do a special field experiment to compare two or more alternative methods of doing a particular job (e.g. the use of combine harvester to harvest a grain-crop may be compared with the use of a binder to cut the crop with later threshing by hand or machine).

Or you may be able to assess a possible change in your methods by testing it in isolation. For example, you can find out by chemical analysis whether hand-mixing or tumbling in a churn gives a more nearly uniform mixture of two fertilisers; this knowledge may enable you to do better fertiliser experiments. Or you may find out by sampling tests whether taking duplicate samples of grain for dry matter percentage would usefully improve the precision of experiments on cereal crops.

In the remainder of this chapter I shall discuss various aspects of techniques and what can be done to assess their effects.

### 9.2 DOWNRIGHT MISTAKES IN APPLYING TREATMENTS

If you interchange two treatments in the same block of a simple randomised block experiment you can usually amend the plan and then forget the mistake — the resulting analysis will be valid. But beware of subtle bias, e.g. has the untreated plot moved into a special place (e.g. the last plot in order)? If you have omitted one treatment and duplicated another, or if you have interchanged treatments between blocks you will probably have to estimate 'missing values' for at least some of the plots subject to mistakes (see Section 13.9).

## 9.3 RANDOM INACCURACIES OF TECHNIQUE

Suppose that the yields you record from the $n$ plots of your experiment are

$$y_1, y_2, \ldots, y_n$$

Suppose also that, if your techniques of sowing, applying treatments, harvesting, etc., had all been perfect the yields then recorded would have been

$$z_1, z_2, \ldots, z_n$$

(we may call these the 'true' yields, as opposed to the 'recorded' yields $y_1$, etc.).

We may write

$$y_1 - z_1 = e_1$$
$$y_2 - z_2 = e_2 \quad \text{etc.,}$$

$e_1, e_2, \ldots, e_n$ being the discrepancies or inaccuracies (I avoid the ambiguous word 'error').

In many (but not all) circumstances we may feel justified in assuming that there is no correlation between the discrepancy $e$ and the true yield of each plot (in particular, that for any value of $z$, $e$ is a random sample from one single population).

If this assumption is correct

variance $(y)$ = variance $(z + e)$ = variance $(z)$ + variance $(e)$,

or, as often written,

$$V(y) = V(z) + V(e)$$

The error mean square of our recorded yields ($E$ say) is an estimate of $V(y)$. If we had used perfect techniques throughout the error mean square we would have recorded may be called $E_p$; this is an estimate of $V(z)$.

It would be nice, then, to write

$$E = E_p + \text{estimate of } V(e)$$

but it wouldn't be true. In the more-or-less small sample offered by a field experiment there is a fair chance that $E$ will be less than $E_p$. This can happen because the values of $e$ in the sample of $n(e_1, e_2, \ldots, e_n)$ may be negatively correlated with the values of $z(z_1, \ldots, z_n)$. When $n$ is large this chance is remote and the equation given above may be true, or reasonably nearly true.

Any randomly-distributed discrepancies introduced into the recorded yields *tend* to increase the error mean square of the yields, but do not necessarily do so in any one experiment. It is often useful to think in terms of percentages; if the SE per plot (with perfect technique) is $s$ per cent, and the discrepancies have an SE per cent of $t$ then, on average over many experiments, the error mean square of the recorded yields will be $s^2 + t^2$.

We can seldom be sure that discrepancies are not correlated with yields, but often it is a useful approximation to assume they are. The rounding error caused by weighing to (say) the nearest kg is probably uncorrelated with yield; the discrepancy caused by small (random) inaccuracies in marking out the harvested area is probably not highly correlated with yield. So this line of thought can be useful.

## 9.4 MINOR INACCURACIES IN RATES OF APPLICATION OF TREATMENT

If in an experiment comparing several rates of a fertiliser the actual amounts applied to each plot differ slightly (in a random way) from the amounts intended two points arise:

(1) The size of the 'error' in the amount will vary with treatment — the 'nil' treatments if there are any will presumably be applied exactly, and as the quantity of fertiliser increases the error will (on average) increase, though less than proportionately. So the variance of different comparisons will be affected in different degrees by the 'errors' of application. In fact many people suspect that even if treatments are perfectly applied the variance of plot-yields increases with mean yield (e.g. on a responsive site plots treated with N have greater variance than those without N) so the effect of 'errors' of application merely enhances an existing difference.

(2) If the distribution of the errors in the rate of application is symmetrical (i.e. if positive and negative errors of equal size are equally common) and the relevant response-curve is of the usual type (i.e. sloping upwards but with gradient decreasing) treatment-effects will be underestimated. See *Figure 9.1*, which also illustrates the fact that a farmer who spreads a given quantity of fertiliser or manure unevenly gets less benefit than he would if he spread the same amount evenly.

## 9.5  TECHNIQUE OF HARVEST

At the moment the experimenter decides the crop is fit to harvest there is a definite amount of the crop on each plot (the 'biological yield', I believe the Russians call it). Of course some of the yield may have been lost already (eaten by birds for instance) and some may be in the wrong place (e.g. grain already shed). But, setting aside these losses, no harvesting method is perfect and what we weigh will not be precisely the 'biological yield'. (It may be less, but occasionally it may be more.) Yield in the wrong place can usually be estimated by doing extra work

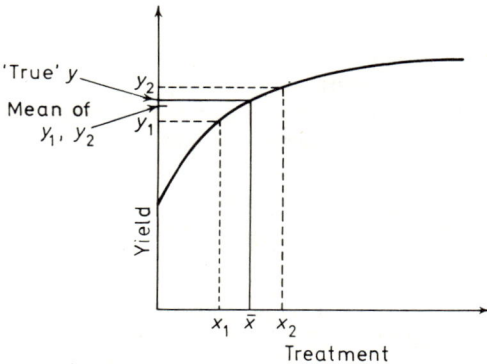

*Figure 9.1*  $\bar{x}$ = intended rate of application; $x_1$, $x_2$ actual rates with equal and opposite errors

on the existing plots; we may pick up shed grain by hand or by vacuum cleaner from sample areas, or we may hand-dig parts of potato plots after mechanical lifting.

But machines such as combine harvesters bring their own problems (even if they are working perfectly). A commercial combine harvester continues to deliver grain at a decreasing rate for several minutes after it has stopped cutting corn. (Special machines can be obtained in which the delivery is completed in a reasonable time but commercial machines are often used on experiments.)

Experimenters allow a fixed 'run-out' time at the end of the cutting of each plot which is a compromise between quick working and complete emptying of the machine. An unknown quantity of grain is carried over from each plot and delivered with the grain of the next plot harvested. This amount may vary systematically; it is probably bigger after the cutting of a plot with a big yield per unit area than after the cutting of a poorer plot. It may also vary in a random way.

It is possible to do a rather artificial experiment to assess these effects directly as mentioned above (Dyke and Meyler[22]). But, if we wish to compare our combine either with another machine or with say, harvesting by hand, it may be worth while to do a field experiment specially for this purpose.


## 9.6  EXPERIMENTS TO COMPARE METHODS

Experiments of this type bring up some new problems which seem to have received rather little attention. The questions we want to ask are usually:

(1) Do the two or more methods give different mean yields when harvesting equal crops?

(2) Do the methods give equal estimates of treatment differences or, put another way, does the difference between any two methods of harvest vary according to the treatment that has been applied to the crop they are harvesting? Or put yet another way, is there a method x treatment interaction?

(3) Are the variances of the 'errors' of the methods different?

I think a little discussion of these three questions is in order. Question (1) deals with the case where one method is regularly less efficient in recovering the crop than another; for example one combine harvester might thresh badly and always leave some grain in the straw. Since a loss of this sort would almost certainly be greater on plots with heavier yields we are led to Question (2) which in this case can be put as 'do losses vary with the treatment of the plot?'. It is possible too that two combine harvesters that give equal estimates of the yield of a crop in good condition may differ in their efficiency when working in a lodged crop, or an over-ripe crop liable to 'shatter' at the slightest touch.

So far so good. Obviously the 'best' method of harvest is one that recovers most of the crop and that is least affected by variations of the state of the crop, if any method exists that combines both superlatives. But, anyway, *is* it the 'best'? Cutting by hand with vacuum-cleaning of the stubble afterwards would probably 'beat' any combine harvester on both counts — c'est magnifique mais ce n'est pas l'agriculture. If we use such an 'ideal' method we are lessening the relevance of our results to farmers' practices. Combines must be accepted, warts and all, if they are commonly used by farmers. I do not mean to imply that hand harvesting is never to be used, but if it is used, this limitation of relevance (or validity) must be remembered.

But what about Question (3)? Clearly if combines A and B are

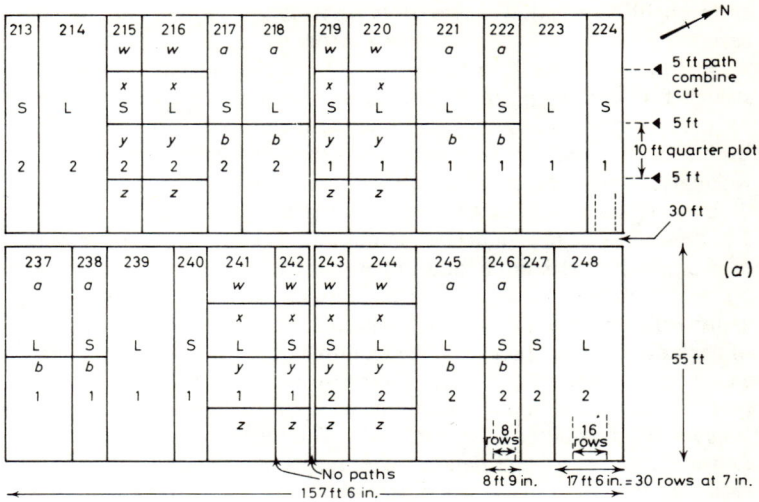

| 213 | 214 | 215 | 216 | 217 | 218 | 219 | 220 | 221 | 222 | 223 | 224 |
| | | w | w | a | a | w | w | a | a | | |
| | | x | x | | | x | x | | | | |
| S | L | S | L | S | L | S | L | L | S | L | S |
| | | y | y | b | b | y | y | b | b | | |
| 2 | 2 | 2 | 2 | 2 | 2 | 1 | 1 | 1 | 1 | 1 | 1 |
| | | z | z | | | z | z | | | | |

N

5 ft path combine cut

5 ft
10 ft quarter plot
5 ft

30 ft

| 237 | 238 | 239 | 240 | 241 | 242 | 243 | 244 | 245 | 246 | 247 | 248 |
| a | a | | | w | w | w | w | a | a | | |
| | | | | x | x | x | x | | | | |
| L | S | L | S | L | S | S | L | L | S | S | L |
| b | b | | | y | y | y | y | b | b | | |
| 1 | 1 | 1 | 1 | 1 | 1 | 2 | 2 | 2 | 2 | 2 | 2 |
| | | | | z | z | z | z | | | | |

(a)

55 ft

8 rows     16 rows

No paths
157 ft 6 in.

8 ft 9 in.    17 ft 6 in.= 30 rows at 7 in.

Treatments

*Nitrogen*
1,2        0.4, 0.8 cwt N per acre

*Combine harvester*
L        Large combine (9 ft 4 in.cut - 16 rows)
S        Small combine (4 ft 8 in.cut - 8 rows)          (b)
*Plot length harvested* (between blank rows)
In quarter plots, half plots, whole plots

*Figure 9.2* A Rothamsted experiment to compare two combine harvesters on barley (half the blocks omitted for clarity). Design: 4 pairs of blocks (N confounded with blocks); each block split into 3 sub-blocks for plot-length harvested, each sub-block split into 2 for size of combine

judged equal according to Questions (1) and (2) but A carries over from plot to plot a very variable amount of grain whilst B carries a constant amount (or none) then B is 'better' than A; if A is used the error mean square of the experiment will be greater than if B is used. To get the same information using A we shall need more replication than if B is used. If the intrinsic error mean square, i.e. the error mean square that would be obtained with a perfect method of harvesting, is really big, then errors of harvesting matter relatively little. But if the coefficient of variation of the harvestable crop is small (say 2 per cent) then even quite small variability in performance of the combine causes a serious loss of information. There is also the nagging thought that errors of this type are probably correlated — two plots harvested in succession almost certainly tend to have compensating errors.

To devise an experiment to answer Questions (1) and (2) is fairly straightforward; we regard 'methods' as one factor and we put in one or more other factors such as varieties of the crop or rates of application of nitrogen to give crops of widely differing characteristics. And we analyse and interpret the results in the usual way.

Devising an experiment to answer Question (3) is another matter. For one thing we need a design that allows us to estimate the variance of each method separately and this implies having at least duplicate plots of each method together in each block — in fact we need an analysable design of plots of one method interlaced with similar plots of each of the other methods. An example of a design that to some extent meets these requirements is shown in *Figure 9.2*. The width of plot varied because the two combines cut different widths and only a single width was taken from each plot. The two machines were tested on full length, half-length and quarter-length plots because it was expected that the smaller machine would retain less grain (and a less variable amount) than the large one and so differences in error variance might be important when harvesting short plots.

Two such experiments were done at Rothamsted in 1967, one on wheat, one on barley (Dyke[22]). Although the results of the two differed considerably, they suggested that:

(1) the large combine underestimated yield on the shortest units; the smaller one gave equal yields from all lengths;
(2) the small combine gave less variable yields than the large on units of all three lengths — although being narrower it was harvesting a smaller area.

# Chapter Ten

# HISTORICAL NOTES ON FIELD METHODS

## 10.1 BEGINNINGS

The early comparative experiments on crops were generally made by dividing a field into two or three more or less equal areas and using these as 'plots' in an unreplicated experiment. From this experimenters progressed to the use of tidy plots marked out carefully and of exactly equal area. This made for easier calculation of yields per acre (and most of the early experiments were done in countries where the unit of area *was* the acre). It also allowed experiments to be sited clear of too much shading by hedgerows, trees and other anomalies likely to occur near field boundaries. Experimenters continued to use large plots, often comparable in size and shape to a tennis court. It was perhaps lucky that the ideas of small plots and multiple replication did not arrive too soon, for the use of large squarish plots in manuring experiments minimised the effects of soil movement and so allowed experiments to be validly continued for many decades.

The idea of including a factorial set of treatments dates back at least to 1852 when Lawes and Gilbert started the Hoos Field Barley Experiment. And in 1856, when they started the Park Grass Experiment they included duplicate unmanured plots; an early example of replication.

By 1910 enough uniformity trials had been done to show the advantages of replicated small plots and to give an idea of the intrinsic variability of typical experimental sites (Mercer and Hall[39]). But experimenters seem to have been reluctant to accept the implications of this work and many unreplicated experiments on large plots were still initiated. The contrasted views are nicely illustrated by the apocryphal remark attributed to a member of the Rothamsted Conference on Field Methods[45]: 'give me one plot and I know where I stand'.

The plant breeders seem to have taken the next step forward. By 1922 they were discussing two contrasted methods of experimentation. One, the 'chessboard' method, involved large numbers of very small plots sown by hand. This allowed valid efficient designs to be used but had the disadvantage that the crop was grown in conditions of soil and spacing that were very different from those of field crops grown by farmers.

The other, the 'half-drill-strip' method is discussed in the next section.

## 10.2 THE HALF-DRILL-STRIP METHOD

This was much used in comparing varieties of cereals about 1920–1930; since then it has almost disappeared. Its drawbacks are partly statistical and partly otherwise and a discussion of them will illustrate the difficulty I feel in trying to give an account of the development of field methods without reference to the simultaneous development of statistical thought.

The method (see Beaven[15]) is to use a wide drill with its seed-hopper divided so that one variety is sown on the left hand side, another on the right. The drill was 'led' (there was a certain politeness in those days of horses; now we 'drive' tractors and drills) up and down – that is, neighbouring widths were drilled in opposite directions. 'Joins' were made as in ordinary non-experimental drilling, often by guiding one wheel of the drill along the mark made by the same wheel on the last width. The result was a pattern of strips in which the two varieties alternated. Each strip (except for one at each end) was one drill-width wide. Beaven advocated a minimum of 10 strips of each variety, each made up of two half-drill widths sown in successive passes of the drill; this implies 11 passes of the drill in all. The odd half-drill strip at each end was presumably discarded. Narrow paths could be left between varieties by shifting all the coulters of the drill a little out from the centre-line. Alternatively, edge-effects could be lessened by sowing an 'interference strip' a few rows wide between the varieties.

This method had two great advantages. First, drilling was done as in a farm crop, the drill at each pass being guided by the wheelings of the previous pass. Thus, once the drill was suitably modified and filled an experiment could be drilled by farm staff with very little super-vision; in particular, nothing needed to be marked out except the four sides of the area (strictly, three sides were enough, the fourth merely telling the drillsman when to stop!).

Secondly, since the two varieties were sown simultaneously any change in conditions, e.g. a rainstorm making the soil sticky and spoiling the action of the drill or even interrupting the work altogether, affected both varieties equally. A slow change, such as the gradual drying of a seedbed, also affected each equally.

The disadvantages of the half-drill-strip method are, however, grave. First, the number of varieties that could be accommodated in any one 'experiment' was two. If more varieties were to be compared it was usual to designate one as 'control' and make separate comparisons between it and each of the remainder. (A modern experimenter would probably eschew this wasteful procedure and pair the varieties by using a balanced design in incomplete 'blocks' of two varieties each.) If each drill-width strip is counted as a plot, the efficiency of either arrangement is likely to be very small in comparison with the modern arrangement in randomised blocks each containing one plot of each variety (unless standard error per plot increases very greatly with number of plots per block, i.e. unless yields of neighbouring plots are closely correlated).

Secondly, we can never be sure the various coulters of a drill behave identically; very often they clearly do not. Some variation between coulters is systematic (e.g. between those that follow the tractor-wheels and those that don't) and some is 'random' but consistent from length to length sown. ('Random' in this context means 'unknown' or 'not allowed for' — as indeed it does in many contexts.)

The systematic variation (in so far as it is symmetrical about the centre line) will not affect the comparison of varieties. But the random component of variation between coulters tends to increase the yields of one variety relative to the other. If an attempt is made to estimate 'error' variance by comparing differences between successive pairs of strips the random component will not be allowed for — the error will be 'biased downwards'.

But there is another reason why the estimate of error based on 'varieties x strip-pairs' is invalid, a more subtle reason which illustrates a fundamental point of 'the logic of uncertain inference'. In the methods described above there is only one act of randomisation that the experimenter can perform. He can choose to put variety A at the extreme (say) East of the experiment; or he can put variety B there. Once the decision is made the rest of the layout is settled and barring mistakes cannot be altered. Logically he is in the same situation as if he had merely two large plots side by side (as his predecessors had). The fact that each plot has been replaced by a number of narrow strips with the intervening spaces constituting the other 'plot' may increase the precision of the comparison but it does nothing to provide a valid estimate of error. A valid estimate of error can be made if and only if successive acts of randomisation have been made independently. It may help if you think of a site where there are fertility 'waves' so that under uniform treatment the yield of every even strip (2nd, 4th, 6th, . . ., from the East) is greater than those of the adjacent strips. For clarity, imagine the yields are

$$10 \quad 20 \quad 10 \quad 20 \quad 10 \quad 20 \ldots$$

and that we are going to use them to compare two varieties A and B which happen to give identical yields under all conditions. If we use the half-drill strip method there are just two possible results:

(1) mean of A = 10
    mean of B = 20
(2) mean of A = 20
    mean of B = 10

and in each case (judging by varieties x strip-pairs) the error variance is zero.

If we use randomised blocks of two plots, arranged in the obvious way, the extreme results given above *may* occur — but each is extremely unlikely. We are much more likely to get a result such as

mean A = 12

mean B = 18

with a non-zero estimate of error variance. We have an analysable design with a valid estimate of error. We've given the design a pretty rough handling and it's under strain but its logical basis is unimpaired.

I have of course (as I so often do) given an absurdly extreme example as an illustration. But remember that a complicated pattern of variation in yield across a site can be resolved into components (just as the jumbled waves of the sea can be sorted out according to their wave-lengths) and a site with a strong component of variation with a 'wave-length' of the same order as a drill-width is not necessarily uncommon.

I conclude this chapter by drawing your attention to five papers that excellently illustrate the development of field methods, and the accompanying statistical thought, up to 1939.

Engledow and Yule[26] have an eminently quotable common sense: 'Picking the best variety for a locality is like picking a horse or a wife — every man must decide what he wants and then set out to get it.' 'The grower measures quality by what he is paid for it.' But I find my last quotation rather severe: 'Yield figures should be accorded no credence, nor even be divulged, unless accompanied by evidence of statistical significance.'

'Student'[47], writing in 1926, when the half-drill-strip method was still in favour, set out clearly and succinctly the logical basis of randomised designs. This paper, written nearly 50 years ago by a brewer, could hardly be bettered by anyone now wishing to write an introduction to the statistics of field experiments.

Fisher[28] in the same year (but in a most improbable journal) gave his arguments for replication and randomisation. Crowther[20] summarised the methods of planning and interpreting experiments in the 1930s; this paper includes an excellent historical section and lists references to early work in Britain, Germany and the U.S.A.

Moffatt[40] gives the fullest account of field methods developed at Rothamsted and elsewhere before combine harvesters were available.

# Part II

Statistics in Field Experiments

In the following chapters I shall not try to give you a course in statistical theory. I intend merely to put before you some of the basic ideas behind the paraphernalia of analysis of variance, tests of significance, etc., and to give you a few different ways of looking at the algebraic and arithmetic operations that are needed in designing experiments and in interpreting their results. There are plenty (perhaps embarrassingly plenty) of books that will give you the formulae to be used in particular cases and you will need to use at least some of these books. I hope these chapters will help you to 'have a feeling' for what you are doing and so lessen the risk that you will do something nonsensical and will help you to modify the given method appropriately when cases a bit out of the ordinary turn up. I know from bitter experience that fluent statistical thinking comes hard to me; the ideas contained in the following two chapters would have been useful to me if they'd been presented to me early on.

# Chapter Eleven

# COMPARISONS, DEGREES OF FREEDOM AND 'ERROR'

## 11.1 RANDOMISATION

Suppose we wish to compare four varieties A, B, C and D and we have eight plots available in two rows of four.

I am sure you know the risk of using a plan such as

|   |   |   |   |
|---|---|---|---|
| A | B | C | D |
| A | B | C | D |

or

|   |   |   |   |
|---|---|---|---|
| A | B | C | D |
| D | C | B | A |

and you will (I hope) agree with me that it will be wise to use 'two randomised blocks of four plots each'. But here are a few points to bear in mind.

(1) *We could* randomise even more and put all eight treatments (two of A, two of B, two of C, and two of D) into one randomised block of eight plots. There are two cases where this is clearly the best course:

(*a*) when we expect to apply another factor at two levels (perhaps nitrogen fertiliser at single and double rate) later. This is a trivial point in an experiment planned to last one year but may be important at the beginning of a long-term experiment.

(*b*) When (i) we are sure that if all plots receive the same treatment the mean differences between blocks of four

101

are little or no greater on average than mean differences between *any* groups of four plots, e.g. the difference represented by the pattern

$$+ \quad - \quad + \quad -$$
$$- \quad + \quad - \quad +$$

and (ii) we need every possible degree of freedom for the estimation of error (the 'fully-randomised' design has four degrees of freedom for error; two randomised blocks of four have three).

(2) The plans shown above are of course valid randomisations — that is, if we randomise honestly we must accept that such repellent patterns *will* turn up occasionally.

There is mental strife (perhaps moral strife) within and between experimenters whether it is ever (or always) right to reject such patterns when they do turn up. Writing personally, I am against abortion, and I have at times accepted patterns quite as extreme as the examples. But I allow other people their own opinions.

In certain cases you can avoid some of the more extreme patterns without doing violence to the sacred principle of randomisation (Grundy and Healy[31] and Dyke[23]).

A point worth remembering is that randomisation produces a 'valid' experiment (in a sense to be discussed later) whatever the pattern of inherent variability of the plots. If the fertility of the site varies so as to give (with uniform treatment) yields such as:

$$10 \quad 10 \quad 10 \quad 20$$
$$10 \quad 10 \quad 10 \quad 50$$

an experiment in two randomised blocks of four is 'valid' though obviously not very valuable. (In this chapter I use the word 'fertility' to indicate the total effect on crop-yield of *all* conditions on a particular plot — soil, diseases, pests, cultivations. Plots are of equal fertility if with uniform treatment they give equal yields of the experimental crop.)

'Validity' is the property that, if you imagine the experiment 'repeated' on the same plots in identical conditions once with every possible randomisation pattern then,

(1) on average the difference between the means of any two treatments is the same as it would have been in each experiment

if there had been no differences in fertility between the plots, or, briefly, estimates of treatment-effects are unbiased;
(2) on average the error mean square does correctly represent the variation between the different sets of estimates of treatment differences (one set from each randomisation), or, briefly, estimates of error are unbiased.

Here is a simple example. You will realise that any more complicated one would need intolerably long discussion but the principle would be identical.

Consider a uniformity trial on four plots, numbered 1 to 4. Suppose the arrangement and yields of crop are:

| Plot No. | 1 | 2 | 3 | 4 |
|---|---|---|---|---|
| Yield | 96 | 92 | 107 | 105 |

(I have for convenience chosen yields whose mean is 100).

We may regard each yield as made up of two parts, the mean of all four yields (the 'grand mean', in this case 100), together with (i.e. added to) a deviation from this mean caused by the fertility of the particular plot, as shown in *Table 11.1*.

**Table 11.1** A UNIFORMITY TRIAL

| Plot | 1 | 2 | 3 | 4 | |
|---|---|---|---|---|---|
| Grand mean | 100 | 100 | 100 | 100 | |
| Deviation | −4 | −8 | +7 | +5 | (Mean = 0 naturally) |
| Total | 96 | 92 | 107 | 105 | |

Next we imagine our four plots grouped into two blocks of two plots each, plots 1 and 2 forming block I, plots 3 and 4 block II (it is common practice to use Roman numerals for block numbers).

Now we regard each yield as made up of three parts, grand mean, deviation of block mean from grand mean, and deviation of plot-yield from block mean. In this case the block means are: block I: 94, block II: 106, and we proceed as in *Table 11.2*.

What we have done so far illustrates in a simple form the idea of 'degrees of freedom'. The original four yields chosen could equally well have been any other four numbers (non-negative of course) I had happened to think of (the mean need not be 100). In other words I had four degrees of freedom when I chose them. The deviations from the grand mean (−4, −8, +7, +5) *cannot* all be chosen independently, since their mean must be 0. Three of them can be given any values (positive or negative) but then there is only one possible value for the fourth. We say 'there are three degrees of freedom for deviations from the mean'.

Next, the two block deviations (−6, 6) involve only one degree of freedom (the choice of one determines the other) and within each

**Table 11.2** THE UNIFORMITY TRIAL AGAIN

| Block | I | | II | | |
|---|---|---|---|---|---|
| Plot | 1 | 2 | 3 | 4 | |
| Grand mean | 100 | 100 | 100 | 100 | |
| Block deviation | −6 | −6 | +6 | +6 | (Mean 0) |
| Plot deviation | +2 | −2 | +1 | −1 | (Mean 0 for each block) |
| Total | 96 | 92 | 107 | 105 | |

block only one of the two plot deviations can be chosen arbitrarily, so there are two degrees of freedom for deviations of plots from block means.

Now, still with the same four plots in mind, we imagine a simple experiment. We are to compare two varieties A and B, using two plots for each. We choose to use two randomised blocks of two plots each. In this case there are only four possible randomisations:

| Block | I | | II | |
|---|---|---|---|---|
| Plot | 1 | 2 | 3 | 4 |
| (i) | A | B | A | B |
| (ii) | A | B | B | A |
| (iii) | B | A | A | B |
| (iv) | B | A | B | A |

Suppose we got (iii) and that, in the conditions of our experiment variety A yields 20 units more than variety B. We assume for the moment that the difference in potential yield between the two varieties is the same at all points of the site, so that if variety A is sown on all plots the yield of each plot will be exactly 20 units greater than if variety B had been sown. (What happens if this assumption is false will be mentioned in Section 11.7.)

Referring back to our uniformity trial (which we suppose was done with variety B) we can calculate the yields we may expect as in *Table 11.3*.

**Table 11.3** AN EXPERIMENT

| Plot | 1 | 2 | 3 | 4 | |
|---|---|---|---|---|---|
| Variety | B | A | A | B | |
| 'True yield' | 100 | 120 | 120 | 100 | (Mean 110) |
| Plot deviation | −4 | −8 | +7 | +5 | |
| Total | 96 | 112 | 127 | 105 | |

Another way of putting this is given in *Table 11.4*.

**Table 11.4** THE EXPERIMENT AGAIN

| Block | I | | II | |
|---|---|---|---|---|
| Plot | 1 | 2 | 3 | 4 |
| Variety | B | A | A | B |
| Grand mean | 110 | 110 | 110 | 110 |
| Block deviation | −6 | −6 | +6 | +6 |
| Variety deviation | −10 | +10 | +10 | −10 |
| Plot deviation | +2 | −2 | +1 | −1 |
| Total | 96 | 112 | 127 | 105 |

Now instead of constructing the example in this way, starting with a sort of divine knowledge of 'true' variety deviation and plot deviations, suppose that we have done the experiment and got these yields:

| Plot | 1 | 2 | 3 | 4 |
|---|---|---|---|---|
| Variety | B | A | A | B |
| Yield | 96 | 112 | 127 | 105 |

and are trying to interpret them. We can calculate the grand mean (110) and the block means: block I 104, block II 116; block deviations −6, +6. (Note that the divine idea of block deviations is identical with the human idea of them in this case; this is not true in experiments with confounding.)

The mean yield of variety A is $\frac{1}{2}(112 + 127) = 119.5$ and that of variety B is 100.5. The deviations are: A + 9.5, B − 9.5. These are not the same as the 'true' values. The reason why variety deviations 'estimated' from the experiment are different from the 'true' ones, but the block deviations, estimated and true, are identical is that varieties A and B are subject to randomisation, blocks I and II are fixed and not subject to randomisation.

We now have the analysis shown in *Table 11.5*.

**Table 11.5**  THE EXPERIMENT ANALYSED

| Block | I | | II | |
|---|---|---|---|---|
| Plot | 1 | 2 | 3 | 4 |
| Variety | B | A | A | B |
| Grand mean | 110 | 110 | 110 | 110 |
| Block deviation | −6 | −6 | +6 | +6 |
| Estimated variety deviation | −9.5 | +9.5 | +9.5 | −9.5 |
| Estimated plot deviation | +1.5 | −1.5 | +1.5 | −1.5 |
| Total (i.e. actual yield) | 96 | 112 | 127 | 105 |

The estimated plot deviations are calculated so as to make the total for each plot equal to the recorded yield of that plot.

## 11.2  THE DEMON ERROR

*Table 11.4* differs from *Table 11.5*. The grand mean is of course the same in both tables, and so are the block deviations (as noted above). But we have been unable to apportion the deviations from block means correctly into variety deviations and plot deviations. Variety A has by chance been sown on two plots whose mean plot deviation (see *Table 11.2*) is −0.5 (and B on plots of mean deviation +0.5). Our reconstruction has amalgamated these contributions with the 'true' variety deviations and we have underestimated the true variety deviations in consequence.

This illustrates the statistical term 'error' (which does *not* imply someone has made a mistake!). 'Error' refers to the discrepancies between true and estimated effects on yields that arise from the unknown deviations of plot-yields from block means. (The definition of 'error' in designs like latin squares where row deviations and column deviations are taken into account instead of only block deviations is a little more complex but not different in principle.)

## 11.3 RANDOMISATION TO THE RESCUE

Consider what would have happened if everything had been the same as in the last two sections except that we had had one of the other three possible randomisations. If we perform for each the same processes of construction (of plot yields) and analysis (giving estimated variety and plot deviations) we obtain *Table 11.6* [repeating case (iii) for completeness].

**Table 11.6** FOUR RANDOMISATIONS ON THE SAME PLOTS

| | Plot | 1 | 2 | 3 | 4 |
|---|---|---|---|---|---|
| (i) | Variety | A | B | A | B |
| | Yield | 116 | 92 | 127 | 105 |
| | Grand mean | 110 | 110 | 110 | 110 |
| | Block deviation | −6 | −6 | +6 | +6 |
| | Estimated variety deviation | +11.5 | −11.5 | +11.5 | −11.5 |
| | Estimated plot deviation | +0.5 | −0.5 | −0.5 | +0.5 |
| (ii) | Variety | A | B | B | A |
| | Yield | 116 | 92 | 107 | 125 |
| | Grand mean | 110 | 110 | 110 | 110 |
| | Block deviation | −6 | −6 | +6 | +6 |
| | Estimated variety deviation | +10.5 | −10.5 | −10.5 | +10.5 |
| | Estimated plot deviation | +1.5 | −1.5 | +1.5 | −1.5 |
| (iii) | Variety (as Table 11.5) | B | A | A | B |
| | Yield | 96 | 112 | 127 | 105 |
| | Grand mean | 110 | 110 | 110 | 110 |
| | Block deviation | −6 | −6 | +6 | +6 |
| | Estimated variety deviation | −9.5 | +9.5 | +9.5 | −9.5 |
| | Estimated plot deviation | +1.5 | −1.5 | +1.5 | −1.5 |
| (iv) | Variety | B | A | B | A |
| | Yield | 96 | 112 | 107 | 125 |
| | Grand mean | 110 | 110 | 110 | 110 |
| | Block deviation | −6 | −6 | +6 | +6 |
| | Estimated variety deviation | −8.5 | +8.5 | −8.5 | +8.5 |
| | Estimated plot deviation | +0.5 | −0.5 | −0.5 | +0.5 |

Note, first, that each of the four randomisations leads us to a set of estimated plot deviations that are (two positive, two negative) of the same magnitude; this corresponds to the fact that there is only one

degree of freedom for 'error' — another way of looking at this fact will
be given in the next section.

Next, let us assemble the four estimates of the deviation for variety
A (the ones for variety B are equal in magnitude but of opposite sign),
see *Table 11.7*.

Table 11.7   ESTIMATED DEVIATIONS OF VARIETY A

| Randomisation | |
|:---:|:---:|
| (i)   | +11.5 |
| (ii)  | +10.5 |
| (iii) | +9.5 |
| (iv)  | +8.5 |
| Mean  | +10 |

The mean of the estimates is exactly equal to the 'true' deviation of
variety A. This illustrates one of the fundamental virtues of all correctly
randomised designs: that the estimate of each treatment effect is
unbiased, i.e. the average value over all possible randomisations is equal
to the true value.

We now come to the second great virtue of correctly randomised
designs — the provision of an unbiased estimate of error.

Returning to our simple example, we tabulate the differences
between the four estimates of the deviation of variety A and the true
value (+ 10). Side by side with these I put the estimated deviation for
plot 1 (*Table 11.8*).

Table 11.8   VARIETY DEVIATIONS AND PLOT DEVIATIONS

| Randomisation | Variety A estimate − true | Plot deviation |
|:---:|:---:|:---:|
| (i)   | +1.5 | +0.5 |
| (ii)  | +0.5 | +1.5 |
| (iii) | −0.5 | +1.5 |
| (iv)  | −1.5 | +0.5 |

If we were to do the corresponding thing for variety B and the
other three plots we should get the same pattern of figures but with
different arrangements of + and − signs. (Note at this point that none
of the estimates of the deviation of plot 1 is equal to the 'true' deviation,
which is +2. This is because in any randomisation our estimate of
deviation of each variety includes the mean of the deviations of the
plots on which that variety is sown, and this inevitably leaves a

different set of residual deviations unless by chance the mean of the original deviations is zero. You may find this clearer if you imagine that A and B are 'truly' equal; our estimated deviations for variety A is then equal to the mean deviation of the A plots.)

Evidently in this very simple case the two numbers 1.5, 0.5 (ignoring signs) have an important role. In randomisations (ii) and (iii) which by chance have given the more accurate estimates of the variety deviations the estimated plot deviations are large; in (i) and (iv) where the estimates of the variety deviations are less accurate, the estimated plot deviations are smaller. This illustrates an important fact: what you gain on the swings you lose on the roundabouts, or, more precisely, a randomisation that by chance gives particularly accurate estimates of treatment differences will give an error mean square greater than average, and so will be judged less accurate than it really is. Vice versa, a randomisation that gives relatively inaccurate estimates of treatment differences will give a small error mean square and will be judged more accurate than it is. Never fear, though, with plenty of degrees of freedom for error this type of misjudgment is rarely important.

Before we can generalise to more complex experiments we need some new ideas, which will be introduced now.

## 11.4 CONTRASTS AND DEGREES OF FREEDOM

We noted that, in the analysis of any one randomisation of our four-plot experiment the estimated plot deviations all have the same magnitude, two positive, two negative. A little algebra will show that each is obtained from the same formula (apart from the change of sign). In randomisation (iii) the deviation of plot 1 is (in terms of plot yields)

$$\tfrac{1}{4}(+96-112+127-105)$$

and so is that of plot 3; those of plots 2 and 4 are each equal to

$$\tfrac{1}{4}(-96+112-127-105)$$

Similarly the block deviations are

$$\tfrac{1}{4}(+96+112-127-105) \text{ for plots 1 and 2}$$

and

$$\tfrac{1}{4}(-96-112+127+105) \text{ for plots 3 and 4}$$

Finally the variety deviations are

$$\tfrac{1}{4}(-96+112+127-105) \text{ for variety A}$$

and

$$\tfrac{1}{4}(+96-112-127+105) \text{ for variety B}$$

If for the moment we ignore the signs and the factor $\frac{1}{4}$, the three formulae

$$96-112+127-105$$

$$96+112-127-105$$

$$96-112-127+105$$

are the only possible ones involving the four plot yields, two positively and two negatively. These, together with the total

$$96+112+127+105$$

contain all the information provided by the original four yields; the four plot yields can be reconstructed if the values of the four formulae are known.

It is customary to use the factor $\frac{1}{2}$ with the first three (but see Glossary entry for 'effect'), so that each is the difference between the mean of two plots and the mean of two others, and so gives a numerical value to the 'contrast' or 'comparison' between the two pairs of plots concerned. Similarly in more complicated experiments we may make comparisons between unequal sets of plots, e.g. the formula

$$\tfrac{1}{2}(y_1 + y_2) - \tfrac{1}{3}(y_3 + y_4 + y_5)$$

($y_1, y_2, \ldots, y_5$ being the yields of five plots), represents the contrast between the mean of two plots and the mean of three other plots.

Whenever we think of a degree of freedom it is possible to write down a formula that expresses the corresponding contrast. When we think of a set of $n$ degrees of freedom it is possible to write down $n$ formulae which express the corresponding contrasts. There are two snags here. First, there are alternative sets of contrasts corresponding to one set of degrees of freedom. For example the two degrees of freedom for comparisons between three yields $y_1$, $y_2$, $y_3$ can be written down as

$$y_1 - y_2$$

and

$$\tfrac{1}{2}(y_1 + y_2) - y_3$$

or, equally validly, as

$$y_1 - y_3$$

and

$$\tfrac{1}{2}(y_1 + y_3) - y_2$$

Second, for all practical purposes, it is essential that the contrasts in one set should be 'mutually orthogonal'. This is why, in the case just mentioned, I did *not* suggest

$$y_1 - y_2$$

and

$$y_1 - y_3$$

as a suitable pair of degrees of freedom.

Although for most purposes, we can use expressions such as $(y_1 - y_2 + y_3 - y_4)$ to define contrasts, or put in the appropriate divisor, e.g. $\frac{1}{2}(y_1 - y_2 + y_3 - y_4)$ and call the expression a 'main effect' or inter-action, when we come to summing squares (as we soon shall) we need a special divisor. This is the square root of the sum of squares of the coefficients of the plot-yields involved in the contrast. For $(y_1 - y_2 + y_3 - y_4)$ the divisor is $\sqrt{[1^2 + (-1)^2 + 1^2 + (-1)^2]} = \sqrt{4} = 2$, so our 'main effect' is acceptable. For $(y_1 - y_2)$ however, the divisor is $\sqrt{2}$. Putting in this divisor can be called 'normalising' the contrast. Any contrast between $n$ yields $y_1, y_2, \ldots, y_n$ can be written as $l_1 y_1 + l_2 y_2 + \ldots + l_n y_n$, in which each of the numbers $l_1, l_2, \ldots, l_n$ may have any value (positive, negative or zero) provided the condition

$$l_1 + l_2 + \ldots + l_n = 0$$

is satisfied. The normalised version of the contrast $l_1 y_1 + l_2 y_2 + \ldots + l_n y_n$ is $(l_1 y_1 + l_2 y_2 + \ldots + l_n y_n)/\mathrm{N}(l_1^2 + l_2^2 + \ldots + l_n^2)$ and, for example, the normalised version of $\frac{1}{2}(y_1 + y_2) - y_3$ is $(y_1 + y_2 - 2y_3)\sqrt{6}$.

## 11.5 ORTHOGONALITY (see also Chapter 12)

Let us go back (once more) to our four-plot experiment. If instead of numbers we use symbols

$$y_{a1} = \text{yield of variety A in block I}$$

$$y_{b1} = \text{yield of variety B in block I}$$

and so on, then the grand mean (usually denoted by $\bar{y}$) $= \frac{1}{4}(y_{a1} + y_{b1} + y_{a2} + y_{b2})$, the variety-contrast $(V) = \frac{1}{2}(y_{a1} - y_{b1} + y_{a2} - y_{b2})$, the block-contrast $(B)^* = \frac{1}{2}(y_{a1} + y_{b1} - y_{a2} - y_{b2})$, and the error-contrast $(E) = \frac{1}{2}(y_{a1} - y_{b1} - y_{a2} + y_{b2})$. Now consider what would happen if the yields of variety B were increased by, say, 10 units.

$$V \text{ becomes old } V - 10$$

but $E$ and $B$ are unchanged ($\bar{y}$ is increased by 5 but this need not

* $B$ for blocks; nothing to do with variety B.

concern us; we are only interested in the contrasts between the four yields at present).

Equally, if we imagine the yields of each plot of variety A changed by 10 units, or any other amount, positive or negative, $B$ and $E$ will be unaltered. We can combine both sorts of disturbance, and if we choose to alter the A yields and the B yields equally but in opposite directions we shall leave $B$, $E$ and $\bar{y}$ all unchanged.

Similarly, if we add an arbitrary amount to each of the yields of the plots of Block I and subtract the same from each of the yields of the plots of Block II, then $V$, $E$ and $\bar{y}$ are unchanged; and finally, if we change the yield of each plot whose $y$ appears positively in contrast $E$, and make the opposite change where a negative sign occurs in $E$, then $B$, $V$ and $\bar{y}$ are unchanged.

This, in a simple case, is the essence of 'orthogonality'. Our contrasts $B$, $V$, $E$ are orthogonal in this case because, if we consider the plots that have + signs in any one contrast these plots have + and − signs in equal numbers in each of the other contrasts, and similarly for those with − signs in the chosen contrast.

Here is the general rule:

If two contrasts between $n$ yields $y_1, y_2, \ldots, y_n$ are written as

$$l_1 y_1 + l_2 y_2 + \ldots + l_n y_n$$

and

$$m_1 y_1 + m_2 y_2 + \ldots + m_n y_n$$

($l_1 \ldots l_n$ and $m_1 \ldots m_n$ being any numbers subject to

$$l_1 + l_2 + \ldots + l_n = 0$$
$$m_1 + m_2 + \ldots + m_n = 0$$

otherwise they are not contrasts) these 2 contrasts are orthogonal if

$$l_1 m_1 + l_2 m_2 + \ldots + l_n m_n = 0$$

With three yields $y_1$, $y_2$, $y_3$ we can use contrasts

$$y_1 - y_2, \text{ i.e. } (1)y_1 + (-1)y_2 + (0)y_3$$

and     $$\tfrac{1}{2}(y_1 + y_2) - y_3, \text{ i.e. } (\tfrac{1}{2})y_1 + (\tfrac{1}{2})y_2 + (-1)y_3$$

and since     $$(1)(\tfrac{1}{2}) + (-1)(\tfrac{1}{2}) + (0)(-1) = 0$$

these are orthogonal. But

$$y_1 - y_2, \text{ i.e. } (1)y_1 + (-1)y_2 + (0)y_3$$

and

$$y_1 - y_3, \text{ i.e. } (1)y_1 + 0(y_2) + (-1)y_3$$

are *not* orthogonal since

$$(1)(1) + (-1)(0) + (0)(-1) = 1, \quad \text{not } 0.$$

A set of contrasts can be said to be orthogonal to another set (and we need not specify those in each set in detail). For example if an experiment has equal numbers of plots with each of the nine combinations of two factors (A and B say) at three levels each ($a_0$, $a_1$, $a_2$ and $b_0$, $b_1$, $b_2$) then the set of (two) contrasts corresponding to factor A are orthogonal to the set corresponding to B; but, as we've seen above, there are several alternative ways of defining the two contrasts for each factor. In this case *any* contrast we choose that involves factor A is orthogonal to any contrast involving factor B.

## 11.6 ESTIMATES OF ERROR IN MORE COMPLICATED EXPERIMENTS

Now we can generalise from our four-plot experiment to the general case of any experiment in randomised blocks (without confounding for the moment).

If there are $t$ degrees of freedom for treatments we can write down $t$ orthogonal contrasts which between them fully specify all treatment differences. If there are $e$ degrees of freedom for error we can (in principle; in some cases it would be an intolerable job) write down $e$ orthogonal contrasts, all representing 'error' (i.e. orthogonal to blocks and treatments). Then, with the same assumption that treatment deviations and plot deviations are additive it is true that:

For any given set of treatment deviations and plot deviations, if we write down the value of every treatment and error contrast (duly normalised) for every possible randomisation, thus:

| Randomisation | Deviation of treatment contrasts | | | | | Error contrasts | | | | |
|---|---|---|---|---|---|---|---|---|---|---|
| | 1 | 2 | . | . | $t$ | 1 | 2 | . | . | $e$ |
| (i) | | | | | | | | | | |
| (ii) | | | | | | | | | | |
| . | | | | | | | | | | |
| . | | | | | | | | | | |

(This is a more complicated table built to the design of *Table 11.8*.)

the sum of the squares of all entries under 'Deviations of treatment contrasts' and the sum of squares of all entries under 'Error contrasts' are in the same ratio as their respective numbers, i.e. $t : e$. In other words, the mean square of the treatment-contrast deviations is equal to the mean square of the error contrasts.

In fact, each is equal to the mean square deviation in a uniformity trial using the same design on the same plots.

In our four-plot example the mean square is 5 which can be derived as

$$\frac{2^2 + (-2)^2 + 1^2 + (-1)^2}{2} \text{ (2 d.f.)}$$

(from the uniformity trial, see *Table 11.2*) and as the mean of $(0.5)^2 + (-0.5)^2 + (-0.5)^2 + (0.5)^2$ (1 d.f.) for randomisation (i) and three expressions similarly obtained from the other randomisations given in *Table 11.6*.

Note that I have not given a proof of this statement; and I have only tried to indicate its plausibility. I have done this at length because I think it important to realise that all this can be done without any assumption about 'distribution of errors'. We have *not* had to postulate any under-lying distribution at all. In fact the analysis of variance and the calcula-tion of standard errors do not depend on the assumption of any particular distribution of errors.

It is common practice to assume a 'normal' or 'Gaussian' distribution of errors for two reasons:

(1) If you take large samples of individuals from almost any 'reasonable' distribution the mean values derived one from each sample are very nearly 'normally' distributed. (And, after all a plot-yield is the mean of a lot of individual plant-yields, at any rate in many crops.)
(2) It allows the valid use of significance tests based on the tables of Fisher's $z$ (equivalent to $F$ which is easier to use) and 'Student's' $t$

## 11.7 NON-HOMOGENEITY OF ERRORS

At the beginning of the discussion we assumed that the plot deviation of a particular plot was unaltered whatever treatment was applied to the plot, i.e. that plot deviations and treatment deviations were additive. They often are not, and we must consider some of the consequences.

First, we must get one thing clear: if we first imagine that variety A is grown on all plots of an experiment (I stay with the idea of a variety trial for simplicity; for a different type of experiment, e.g. a factorial one, the argument may be more complicated but the principle is the same) the deviations of the plots, measured from block-means, have a total of zero — by definition. If we now imagine variety B grown on all plots every individual plot deviation may be changed from its former value, but the total must again be zero — otherwise our idea of variety deviation is incorrect and it must be adjusted accordingly. The block deviations will in the general case, when plot deviations and treatment deviations are not additive, have changed from their original values. And the sum of squares of the plot deviations will in general have a new value.

Next, in any experiment with only two treatments (such as our simple example) it does not matter at all if plot deviations and treatment deviations are not additive. We are estimating one difference (the difference between treatment A and treatment B) and our mean estimate is made up from individual comparisons, one per block, between pairs of plots. Each is subject to an error compounded of an A-plot deviation and a B-plot deviation, which is as it should be. The error assigned to our mean estimate is appropriately based on the sum of squares for differences between these individual estimates.

When, however, more than two treatments are being compared in one experiment we must be on guard against non-homogeneous errors. For example, suppose an experiment has three levels of N-fertiliser, 0, 1 and 2, and that the mean yields are

$$N0 : 10$$
$$N1 : 30$$
$$N2 : 35$$

It is very likely that the plot deviations on plots of treatments N1 and N2 are greater than those of treatment N0. The result is that (if the design permits) we may wish to calculate different standard errors, one for the mean of N0, another for each of the means of N1 and N2 (the difference in variability of these is likely to be negligible).

If the experiment is simply so many blocks of three (each block containing one plot of each treatment) we cannot do this directly (since for N0 we have so many blocks of one plot each).

But we can calculate an error mean square for the contrast

$$N0 - (N1 + N2)/2$$

and a separate one for $(N1 - N2)$ and use these in interpreting our yields.

If the experiment is factorial, e.g. (N0, N1, N2) x (P0, P1, P2) with suitable replication we can calculate separate standard errors for all plots with N0, all with N1 and all with N2 directly, using the formal interactions

P x blocks calculated separately for N0 plots, for N1 plots, and for N2 plots

Experiments sometimes turn up in which a nutrient has been tested at several rates on a site where the soil is extremely deficient in this nutrient. If the results are analysed uncritically you may find the standard error per plot is larger than the yields of some or all of the untreated plots. This is patent nonsense, as it implies (if a reasonable distribution of plot deviations is assumed) that negative yields might reasonably have turned up on some plots. This is a red light that must not be ignored.

Non-homogeneous sums of squares for error may sometimes occur when an experiment has one or more treatments replicated within each block. For example, in experiments comparing say three forms of N-fertiliser it is common to include triplicate plots without N in each block; if there are three rates (including zero) the experiment may be regarded as a 3 x 3 with dummy comparisons, since at rate zero the factor 'forms' is without meaning. In such experiments it is possible to make a separate estimate of the error mean square by using the sum of the squares of the differences between the replicate plots within each block. This can be compared with the estimate made in the ordinary way from the formal interaction treatments x blocks. In the 3 x 3 case mentioned, with $r$ blocks of nine there are $2r$ degrees of freedom for error based on replicates within blocks and $6(r - 1)$ degrees of freedom for error based on treatments x blocks (there are seven distinct treatments).

As far as I know there can be no reason (except a fluke of the randomisation) for the within-block mean square to exceed the treatments x blocks mean square. I assume your experiment has been well planned; if the discards are inadequate yields may be influenced by *competition effect* and this may inflate the treatments x blocks mean square. Situations can be imagined, however, where the opposite may be expected.

If in one experiment the second estimate is significantly greater than the first, or if in a series of experiments the second often exceeds the first, what do we conclude, and what do we do? We *conclude* that differences in condition from block to block do 'really' interact with treatments, or, if you like, that there is block-scale variability on at least some of the sites that interacts with treatments to a degree that within-block variability does not. We might then start looking for explanations in terms of soil variation, past differences of cropping or manuring or differences of exposure to wind or sun.

What we *do*, very reluctantly and only if it seems inescapable, is to throw away the sum of the squares for error based on within-block replication (which is appropriate only for assessing the within-block variation of yield without N) and retain only the mean square based on the treatments x blocks interaction — which is appropriate for assessing the variability of the treatment differences from block to block of the site. [You might conceivably justify using the within-block estimate of error if the validity you want is for just the area of the experiment and no more (not even the rest of the field). See Section 8.12 for an analogous choice. But within-block error is usually applicable to a subset of treatments (e.g. N0) only.] As far as I know, non-one has found a series of experiments in which the two types of error mean squares differed more than might be expected from random variation. But you may.

# Chapter Twelve

## MULTI-DIMENSIONAL GEOMETRY

### 12.1 APOLOGY FOR PUTTING IN THIS CHAPTER

Now we have encountered this rather exotic word 'orthogonal' (which means 'at right angles' in a generalised sort of way) this is perhaps the moment for a short ramble into multi-dimensional geometry. If this leaves you cold, don't worry. I first met this idea in a course of lectures I attended in 1946 or 1947 and it lay fallow in my mind for several years after that, but every now and then the idea helps, like a ray of daylight in a dark room.

### 12.2 BASIC IDEA

The basic idea is this: the yields of the $n$ plots of an experiment are thought of as the co-ordinates of a point in $n$-dimensional space. If $n = 2$ I can show you the idea fully on paper (*Figure 12.1*). (Deceptively easy so far!)

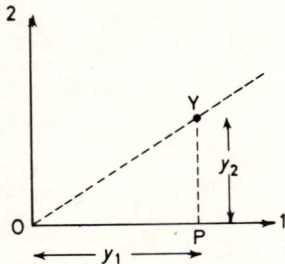

*Figure 12.1*

118

I call the point $Y$ to correspond with the yields $y$ (i.e. $y_1$, $y_2$ in this simple case). $Y$ is usually called the sample point. The origin of co-ordinates is as usual called O. The axes I label with numbers only: 1 for $y_1$ and so on (I want to keep $x$ out of this at present). If $n = 3$ you must use a little spatial imagination (*Figure 12.2*).

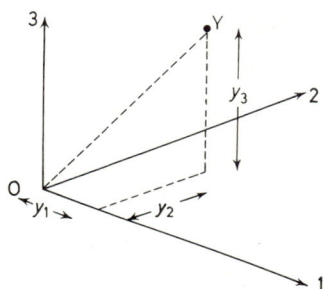

*Figure 12.2*

So far so good, and if I could write profoundly about an experiment with no more than three plots we should all be happy. But even with only three dimensions (i.e. three plots) we *can* do something.

First, if all three yields are equal, i.e. if $y_1 = y_2 = y_3$, the point Y lies on a very special line through O – the line that makes equal angles with each of the axes. May I call this line OM? (M for mean). M may of course be at any distance from the origin (*Figure 12.3*).

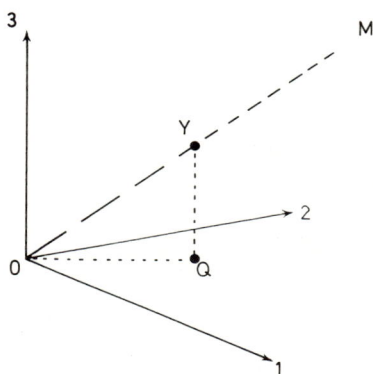

*Figure 12.3*

Note that as the common value of $y_1$, $y_2$ and $y_3$ varies Y moves along OM. Note also that if we drop a perpendicular from Y on to the plane of the axes O1 and O2 the foot of this perpendicular (Q in

*Figure 12.3*) is at equal distances from these two axes. And similarly for the planes of O2, O3 and of O1, O3.

Next if $y_1$, $y_2$ and $y_3$ are *not* all equal, but don't differ very much, Y will lie near to OM and the nearer it is to OM the less the differences between the $y$'s. This is a fairly woolly statement and I want to pull off some of the wool now.

In *Figure 12.1*, by simple application of Pythagoras' theorem,

$$OY^2 = OP^2 + PY^2$$

(if P is the foot of the perpendicular from Y on to O1)

i.e. $$OY^2 = y_1^2 + y_2^2$$

Similarly in *Figure 12.2*

$$OY^2 = y_1^2 + y_2^2 + y_3^2$$

And a similar equation holds in the case of $n$ dimensions:

$$OY^2 = y_1^2 + y_2^2 + \ldots + y_n^2$$

Back to $n = 2$ for the moment. *Figure 12.4* shows what we can do if we plot Y and OM on the same diagram and drop the perpendicular $YY_m$ on to OM. $OY^2$ can be expressed as $OY_m^2 + Y_m Y^2$. We can say the same thing (and a bit more) by saying 'the vector OY has been resolved into orthogonal components $OY_m$ and $Y_m Y$'.

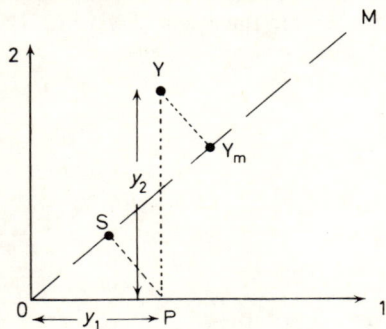

*Figure 12.4*

It is easy to show, since $OY_m = OS + SY_m$ (YP being the perpendicular from Y on to the axis O1 and S the point in OM such that PS is perpendicular to OM; the angles $S\hat{O}P$ and $S\hat{P}Y$ are each $45°$), that

$$OY_m = (1/\sqrt{2})(y_1 + y_2)$$

and that $$Y_m Y = (1/\sqrt{2})(y_2 - y_1) \quad \text{if} \quad y_2 > y_1$$

$$(1/\sqrt{2})(y_1 - y_2) \quad \text{if} \quad y_1 > y_2$$

Something familiar here? Yes, the divisor ($\sqrt{2}$) in each case is the square root of the sum of squares of the coefficients of the $y$'s. $Y_m Y$ is in fact the 'normalised' contrast between $y_1$ and $y_2$. And $OY_m^2 = \frac{1}{2}(y_1 + y_2)^2$ = the correction for mean in the analysis of the variance of the $y$'s. Similarly

$$Y_m Y^2 = \frac{1}{2}(y_1 - y_2)^2$$
$$= y_1^2 + y_2^2 - \frac{1}{2}(y_1 + y_2)^2$$

= sum of squares of deviations of $y_1$, $y_2$ from their mean
(1 degree of freedom)

I am *not* going to give you the corresponding diagram for $n = 3$ — it might put you off $n$-dimensional geometry for life. But I do ask you to agree the thing *can* be extended to $n = 3$, and (in a sort of half-imagination) to $n = 4$, 5 and so on.

When $n = 3$ there is an important difference: the mean is represented by a *line* OM but deviations from the mean are represented by a *plane*. (Imagine the axes O1, O2 and O3 are all the same length; pick them up, tilt them and put them on the table in front of you, O uppermost, to form a triangular pyramid. Then the table is the plane I mean and OM is perpendicular to it — or if you like, the plane is orthogonal to OM.) I'll risk a diagram at this point, *Figure 12.5*; if it looks two-dimensional to you take a short walk and try it again. I have slipped in a point Y (in the plane of the table) to represent the yields $y_1$, $y_2$, $y_3$ of a simple experiment.

In this case note that, if we now forget everything that is not on the surface of the table we have an 'origin' M and a point Y which is in a

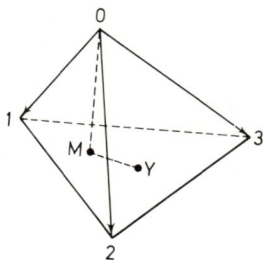

*Figure 12.5*

sense 'free' to move in the plane of the table. We haven't any obvious axes in the plane, but we can use any two lines through M provided they are perpendicular. We could choose M1 and a line parallel to 23 for example. There's an analogy here with choosing two particular contrasts between three yields; we have in effect chosen $(y_2 - y_3)$ and $y_1 - \frac{1}{2}(y_2 + y_3)$.

I hope by now you have realised that dimensions correspond to degrees of freedom. 'Taking out the mean' corresponds to a sum of squares based on 1 d.f. — and so a line (OM) is involved. Two degrees of freedom are left, and, given that $\bar{y}$ (the mean of the $y$'s) is fixed, Y is limited to a plane (the one through $Y_m$ and orthogonal to OM). Note (if it appeals to you) that 'taking out the mean' corresponds to 'projecting the three-dimensional diagram on to the plane perpendicular to OM', just as 'taking out the height' of a house gives a two-dimensional 'plan' of the house.

This is as far as I can take you in three dimensions. If I try to take you into $n$ dimensions we shall be fogged with a hyperplane of $(t - 1)$ dimensions orthogonal to a hyperplane of $(b - 1)$ dimensions, both contained in a hyperplane of $(bt - 1)$ dimensions. (In a desperate attempt to represent an experiment of $b$ blocks of $t$ plots.) But carry away (if you can bear it) the idea that 'taking out' a sum of squares (e.g. for 'treatments') corresponds with collapsing your $n$-dimensional diagram by projection. You can, of course, leave out the mean in your imagined diagram and start with three dimensions all representing contrasts between the (four) yields. But (as we've seen in the case when $n = 3$) there is a lot of arbitrariness about the axes. Things are simpler if they are mutually perpendicular (see Sections 12.5 and 12.6 for the complications that arise if they are not). In fact provided they are mutually perpendicular they can 'be in any directions we like', i.e. O1 may represent $(y_1 - y_2)$ or $(y_1 - y_2 + y_3 - y_4)$ or something horrid like $(y_1 - 2y_2 + 3y_3 - 2y_4)$ (each expression being of necessity orthogonal to $(y_1 + y_2 + y_3 + y_4)$ since we have already 'taken out the mean'). This sort of uncertainty about the axes produces vertigo sometimes but stiffen the sinews and hold on. Before we consider a few applications we need the idea of a frequency distribution in $n$-dimensional space ('$n$-space' for short).

## 12.3  DISTRIBUTION IN $n$-SPACE

The point Y is often called the sample point and an explanation of this phrase is perhaps worthwhile. Forget yields for the moment and consider a sample of $n$ values drawn from a 'reasonable' population.

To be clear, let us think of a Gaussian ('normal') distribution with mean zero. (The argument runs much the same with many other distributions, such as that of the number of heads, less 5, in 10 throws of an unbiased coin, or of the plot-residuals in many field experiments.

The requirements are:

(1) the mean is zero;
(2) the distribution is symmetrical, i.e. positive and negative values of equal size are equally frequent;
(3) larger values are less frequent than smaller ones.)

If you draw a single sample of $n$ values and 'plot' the point Y in $n$-dimensional space using the $n$ values $(y_1, y_2, \ldots, y_n)$ it is reasonable to call Y the sample point. If you draw many samples of $n$ values each you can plot a series of points we may call $Y_1, Y_2 \ldots$ and space, especially near the origin, will begin to be cluttered up with dots. In the case of the coin-throwing example certain points will be 'plotted' many times but there will always be empty space in between. If we use a Gaussian distribution, or plot-residuals measured to umpteen places of decimals, we shall end up with a sort of blob of fog, very dense at the origin and thinning out evenly in all directions; at a great enough distance from the origin the fog is so thin that we cannot perceive it.

You will by now (if you're like me) be thinking in three dimensions only. Well and good. Now imagine yourself looking at your blob of fog from a great distance — for example, from a point far out on the axis corresponding to $y_1$. Take a photograph and examine the (two-dimensional) result. This is a blurred spot and you may interpret the density at any point as the frequency with which the sample point has the appropriate values of $y_2$ and $y_3$, regardless of the value of $y_1$. But instead of looking at the blob from a point on one axis why not look from a point M equidistant from all three axes (in the positive octant)? We are of course back on the line OM used in the last section to represent the mean and the two-dimensional distribution we are looking at is the joint distribution of the deviations from the mean.

Again, this is about the limit in three dimensions. But is it plausible to you now that you might examine a multi-dimensional projection of the original distribution that might represent 'deviations from block-means'? I hope so, at any rate.

But 'block-means' brings us away from distributions with zero mean and back to field experiments, and so to analysis of variance.


## 12.4 ANALYSIS OF VARIANCE — A GEOMETRICAL EXERCISE

This can be considered, geometrically and otherwise, as a process of continually abandoning simpler hypotheses in favour of more complicated ones.

We usually forget the first stage, but it's there all the same:

(1)  Abandon the idea that the mean is zero, look at the blob from a point on OM, consider the distribution of deviations from the mean, take out the mean; all these statements are equivalent.

(2)  Abandon the idea that the resulting $(n-1)$-dimensional distribution is fully symmetrical, look at the blob from a 'direction' corresponding to block-means (in place of OM we unfortunately now have a hyperplane and fog of another sort is probably thickening rapidly), take out 'blocks'; another batch of roughly equivalent statements.

And so on.

At each stage the blob, whether it has $n$ dimensions or less (down to one) has a sort of 'radius' — the mean square deviation from the mean, i.e. the variance of the distribution. If the original hypothesis (of a distribution with mean zero, etc.) is valid then the variance at each stage = the number of dimensions left multiplied by the variance of the original distribution. Remember that in analysing one experiment we have one sample point only; we are imagining the distribution of such points if the experiments were 'repeated' under identical conditions — this is purely notional — see Section 11.3. We estimate the 'radius' by simply measuring the distance between the sample point and the origin O, or the projection of this that's left after taking out the mean, or the mean and block deviations, or whatever corresponds to the particular stage we've reached.

At some stages, e.g. 'taking out blocks', we may notice that the blob has shrunk more than we expected in the process of projection, i.e. its radius has become less in greater proportion than the loss of dimensions would indicate or, in more conventional language, the estimate of residual variance has decreased.

Finally, we may have in addition to the sample point Y (representing the yields of our $n$-plots) another point X whose co-ordinates $x_1, x_2, \ldots, x_n$ represent (say) the pH values of samples of soil from the plots. If OY and OX are close together we say the variates $y$ and $x$ are positively correlated. If OX and OY are almost exactly opposite to each other we say $x$ and $y$ are negatively correlated. Numerically, the correlation coefficient,

$$r = \Sigma(xy)/\sqrt{[(\Sigma x^2)(\Sigma y^2)]}$$

is equal to the cosine of the angle $Y\hat{O}X$.

The formula just given is the correct one if we are still believing in a distribution with mean zero; at the next stage when we have taken out the means (of both $x$ and $y$) the formula is the more familiar

$$r = \Sigma[(x - \bar{x})(y - \bar{y})]/\sqrt{[\Sigma(x - \bar{x})^2 \Sigma(y - \bar{y})^2]}$$

and at later stages more complicated formulae are appropriate.

## 12.5 NON-ORTHOGONAL COMPARISONS

I want to deal with a fairly subtle example in the next section, but first consider an absurdly simple one. Someone is rash enough to do an experiment with this design:

| Block | I | | | II | | |
|---|---|---|---|---|---|---|
| Plot | 1 | 2 | 3 | 4 | 5 | 6 |
| Treatment | A | A | A | B | B | B |

Clearly, we cannot rigidly distinguish between block differences due to fertility and treatment effects. Whatever mean difference we find between the yields of the two treatments a sceptic can object that it may be spurious, being an inherent difference between the blocks.

Geometrically, we see that the line corresponding to 'blocks' is the same as that corresponding to 'treatments' and, however close the sample point Y lies to this line, even if it lies on it (i.e. $y_1 = y_2 = y_3$ and $y_4 = y_5 = y_6$) we cannot say whether Y lies near to one rather than the other.

At the other (desirable) extreme, the lines representing 'blocks' and 'treatments' are at right-angles and we can distinguish clearly whether Y lies particularly close to one or the other. In the next section we consider an intermediate case.

## 12.6  A NON-ORTHOGONAL EXPERIMENT

Consider an experiment in two blocks of three plots thus:

| Block | I | | | II | | |
|---|---|---|---|---|---|---|
| Treatment | $T_1$ | $T_1$ | $T_2$ | $T_1$ | $T_2$ | $T_2$ |
| Yield | $y_1$ | $y_2$ | $y_3$ | $y_4$ | $y_5$ | $y_6$ |

The contrast that represents the difference between blocks (ignoring the disturbing effect of treatments) is

$$B = y_1 + y_2 + y_3 - y_4 - y_5 - y_6$$

Similarly the contrast for 'treatments ignoring blocks' is

$$T = y_1 + y_2 - y_3 + y_4 - y_5 - y_6$$

The sum of products of corresponding coefficients is $+2$ so these are not orthogonal. It is possible to modify $T$ and obtain a contrast $T'$ that *is* orthogonal to $B$ (I shall return to this idea but in different language in Section 13.7) and this leads us to sums of squares for 'blocks ignoring treatments' and 'treatments eliminating blocks' which are additive and so lead to a valid analysis of variance. If we use $B$ and $T$ (without modification) this cannot be done.

The spatial analogy is shown in *Figure 12.6*. Here the axis OB represents blocks (1 d.f.) and OT represents treatments (1 d.f.) and these are *not* perpendicular (I've conveniently omitted dimensions for 'mean' and 'error'). If $YY_b$ is the perpendicular on to OB then $OY_b$ represents 'the most of OY we can explain in terms of blocks, ignoring the treatment direction', i.e. 'blocks ignoring treatments' and $Y_bY$ represents what's left over; if $B'$ is any other point in OB then $YB'$ is longer than $Y_bY$. Does this help you to see why the valid test of significance of treatments in such an experiment is the test of the mean square for 'treatments eliminating blocks' — analogous to $Y_bY^2$?

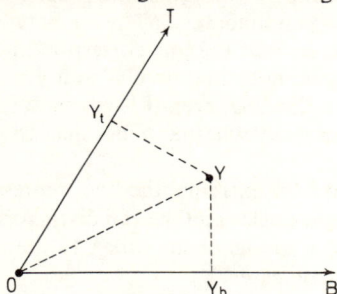

*Figure 12.6*

You can of course drop a perpendicular $YY_t$ on to OT and test the significance of 'blocks eliminating treatments' but this is seldom of much interest.

To conclude this section, let us work out the angle between the axes; six-dimensional geometry can be easier than you think.

OB goes through the point $(1, 1, 1, -1, -1, -1)$ and OT through the point $(1, 1, -1, 1, -1, -1)$. So, if we trace a route from the origin to B taking the co-ordinates in order, 1, 2, 5, 6, 3, 4 the first four steps will serve just as well for a route from O to T. The last two steps are in opposite directions for B and T. After 4 steps we are at a distance $\sqrt{(1^2 + 1^2 + 1^2 + 1^2)} = 2$ from the origin (by repeated use of Pythagoras's theorem). And the remaining two steps for B give an additional distance of $\sqrt{(1^2 + 1^2)} = \sqrt{2}$; for T we have similar steps in opposite directions. So we can boil it all down to *Figure 12.7*. From which we can show by a little trigonometry that

$$\cos \hat{\text{BOT}} = \tfrac{1}{3}$$

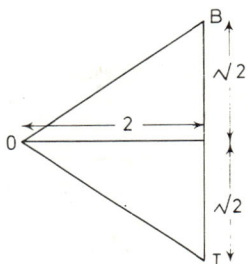

*Figure 12.7*

(The correlation coefficient

$$r = \Sigma(xy)/N[\Sigma x^2 \Sigma y^2] = [(1)(1) + (1)(1) + (1)(-1) + (-1)(1) +$$
$$+ (-1)(-1) + (-1)(-1)]/\sqrt{[(6)(6)]}$$
$$= \tfrac{2}{6}$$
$$= \tfrac{1}{3}$$

which verifies the statement at the end of Section 12.4 for this particular case). So $\hat{\text{BOT}} = 70° 32'$ (about).


## 12.7  TOWARDS COVARIANCE

If you have a 'covariate' $x$ (e.g. the pH of the soil of each plot) the $n$ values $x_1 \ldots x_n$ define a point X just as the yields define the point Y. The length of OX does not matter; its direction is what matters. For a crude and rather inaccurate illustration, imagine yourself at the origin. Y is a star in the sky, X another; if they are 'close together' X and Y are correlated. A test of significance can clearly be worked out by drawing a circle round X of such radius that it contains say 5% of the sky; does Y lie within it or not?

But why not draw a two-dimensional map of the sky by dropping perpendiculars from X, Y, etc., on to the horizontal (or any other) plane? If X', Y' are the feet of the perpendiculars from X, Y then OX' and OY' may be close together even though OX and OY were at a large angle (*Figure 12.8*). Perhaps you would prefer to think of the projections on the plane of the equator of Tromsö and Athens? Now you've found your globe, you can see that the opposite can happen — two sample points (X and Y) that are 'close together' in *n*-space may when

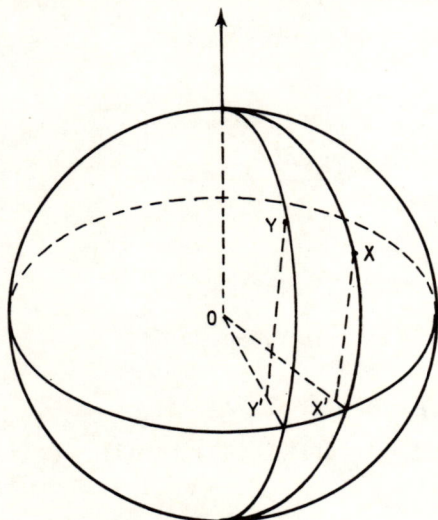

*Figure 12.8*

projected on to a sub-space of less than *n* dimensions not be 'close' — consider two points on opposite sides of the Antarctic continent and then their projections on the equatorial plane. (Their covariance is

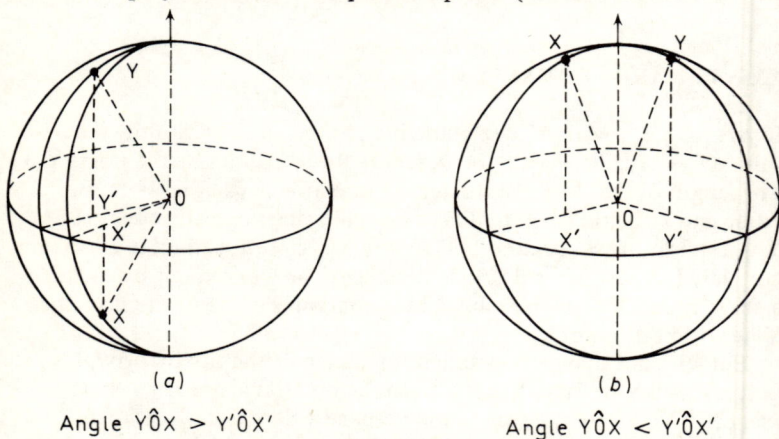

(*a*)

Angle YÔX > Y'ÔX'

(*b*)

Angle YÔX < Y'ÔX'

*Figure 12.9*

positive in three-space, negative in the equatorial two-space; by suitable choice of points in Antarctica you can get positive, negative or zero covariance in two-space.) *Figure 12.9* shows two extreme cases: in (*a*) a small covariance in three-space with a large positive covariance in the equatorial two-space, in (*b*) a large positive covariance in three-space with a negative covariance in three-space. We're at the edge of analysis of covariance now — calculating the regression of $y$ on $x$ after eliminating, say, blocks and treatments.

For precision we must revert to the procedure outlined in Section 12.6. From Y drop a perpendicular $YY_x$ on to OX. Then $OY_x^2$ represents the sum of squares to be ascribed to the regression of $y$ on $x$ and $Y_xY^2$ represents the sum of squares left over. And we can use this procedure in $n$ dimensions (i.e. before taking out the mean, blocks, treatments, etc.) or after 'taking out' all such sources of variation — or at any intermediate stage (i.e. in spaces of assorted numbers of dimensions, all less than $n$).

# Chapter Thirteen

# REGRESSION ANALYSIS

## 13.1 GENERAL

After reading Chapter 11 I hope you are familiar with the idea of making comparisons (in numerical terms) between means of sets of figures, and with the idea of testing the significance of the differences between sets (or combinations of sets).

In this chapter I want to start again, to show you a different approach to analysis — regression analysis — and finally to show the unity that underlies both types of analysis.

I believe Rutherford, in the early days of atomic physics, said that on Mondays, Wednesdays and Fridays he believed light was a stream of corpuscles and on Tuesdays, Thursdays and Saturdays he believed it was a form of wave motion. (I suppose he didn't work on Sundays.) I hope you will adopt a similar attitude to the analysis of experimental data.

We start with a simple example of regression analysis, with one dependent and three independent variates. I assume you know something of the calculation and interpretation of simple linear regressions with a single independent variate. If you don't, consult Rayner[6] or Snedecor and Cochran[9].

## 13.2 MULTIPLE REGRESSION

Suppose that for each of $n$ objects we have measurements $y$, $x_1$, $x_2$ and

$x_3$ (three $x$'s for sake of simplicity, more $x$'s would not affect the argument).

*Note*
  (1) One measurement or 'variate' is distinguished from the rest because we are going to study how $y$ depends on the others. $y$ is the 'dependent variate'; $x_1$, $x_2$ and $x_3$ are 'independent variates' or 'covariates'.
  (2) Strictly there should be a second suffix for each $x$ (and a first one for $y$) to indicate which of the $n$ objects is referred to, e.g. $y_i$, $x_{1i}$, $x_{2i}$, $x_{3i}$ are the measurements for object number $i$, but formulae are simpler to handle without this suffix.
  (3) $n$ must be at least 4, i.e. at least one more than the number of independent variates. Usually the number of objects is much greater than the number of independent variates. For a warning of the pitfalls you may encounter if $n$ is not much greater than the number of independent variates see Section 15.7.

An example is, with $n$ cows,

$$y = \text{milk yield per day of a cow in gallons;}$$
$$x_1 = \text{body weight in kg;}$$
$$x_2 = \text{intake of hay in handfuls;}$$
$$x_3 = \text{length of cow's tail in handspans.}$$

Next we suppose that $y$ is (apart from a random 'error') a linear function of $x_1$, $x_2$ and $x_3$; that is for some numbers $b_1$, $b_2$ and $b_3$

$$y = b_1 x_1 + b_2 x_2 + b_3 x_3 + \text{error ($e$ say)}$$

(This equation really means $n$ equations

$$y_1 = b_1 x_{11} + b_2 x_{21} + b_3 x_{31} + e_1 \text{ and so on)}$$

Now we choose to estimate $b_1$, $b_2$, $b_3$ by making the sum of squares $e_1^2 + e_2^2 + \ldots + e_n^2$ (which I shall call $S$ or $\Sigma e^2$) as small as possible.

*Note*
  (1) This is our chosen method; other methods are possible and (in certain circumstances) better, e.g. if we believe $e$ gets larger as $y$ gets larger.
  (2) I make no assumption about the existence or non-existence of correlations between $x_1$ and $x_2$ (or $x_1$ and $x_3$). It would not matter if $x_2 = x_1^2$. (It would matter if $x_2 = 2x_1$ or if $x_3 = x_1 + x_2$ but this will be discussed in Section 13.6.)

(3) I have not 'taken out the mean'. If you are familiar with regression analysis done on the deviations of each variate from its respective mean, and my presentation worries you, pretend for the remainder of the section that the appropriate subtractions have already been made, i.e. that

$y$ = deviation of milk yield from mean milk yield

and so for $x_1, x_2$ and $x_3$. I hope that Section 13.3 will make you equally at home with either way of working.

We are, then, minimising

$$S = \Sigma(y - b_1 x_1 - b_1 x_1 - b_2 x_2 - b_3 x_3)^2$$

or strictly

$$\sum_{i=1}^{n} (y_i - b_1 x_{1i} - b_2 x_{2i} - b_3 x_{3i})^2$$

regarding this as a numerical function of the 3 unknowns $b_1$, $b_2$ and $b_3$; this leads to equations

$$\left.\begin{aligned}
\Sigma y x_1 &= b_1 \Sigma x_1^2 + b_2 \Sigma x_1 x_2 + b_3 \Sigma x_1 x_3 \\
\Sigma y x_2 &= b_1 \Sigma x_1 x_2 + b_2 \Sigma x_2^2 + b_3 \Sigma x_2 x_3 \\
\Sigma y x_3 &= b_1 \Sigma x_1 x_3 + b_2 \Sigma x_2 x_3 + b_3 \Sigma x_3^2
\end{aligned}\right\} \quad (13.1)$$

These equations can be obtained easily by partial differentiation with respect to $b_1$, then with respect to $b_2$, and finally with respect to $b_3$. Alternatively, a bit of algebra shows that the terms involving $b_1$ in $S$ are:

$$b_1(-2\Sigma y x_1 + 2b_2 \Sigma x_1 x_2 + 2b_3 \Sigma x_1 x_3) + b_1^2(\Sigma x_1^2)$$

Write this for simplicity as $b_1(-2P) + b_1^2 Q$.

$$\text{This} = Q(b_1 - P/Q)^2 - Q(P^2/Q^2)$$

and the value of $b_1$, which minimises this (assuming $b_2$ and $b_3$ are fixed for the moment) is

$$b_1 = P/Q$$

or

$$b_1 \Sigma x_1^2 = \Sigma y x_1 + b_2 \Sigma x_1 x_2 + b_3 \Sigma x_1 x_3$$

(This argument would of course fail if $Q = 0$ but this implies $x_1 = 0$ for every object, a trivial case.)

Now unless there is some linear relation between the $x$'s such as $x_1 = x_2 + x_3$ true for every cow these equations can always be solved for $b_1$, $b_2$ and $b_3$ (and $b_4$, $b_5$, etc., in more complex cases) but the arithmetic can be very heavy.

Now, the minimum value ($S_{min}$) of the sum of squares $S$ is obtained from $\Sigma(y - b_1 x_1 - b_2 x_2 - b_3 x_3)^2$ by putting in the values of $b_1$, $b_2$ and $b_3$ obtained from equations 13.1. Some rather messy algebra shows that

$$S_{min} = \Sigma y - b_1 \Sigma y x_1 - b_2 \Sigma y x_2 - b_3 \Sigma y x_3$$

This equation is very important and should be contemplated long and lovingly. As a sort of verbal rule:

When you fit $b_1$, $b_2$ and $b_3$ you decrease the sum of squares by an amount $b_1$ × (right-hand side of the equation got by considering $b_1$ as the only variable) plus $b_2$ × (right-hand side of the equation got by considering $b_2$ as the only variable) plus $b_3$ × (right-hand side of the equation got by considering $b_3$ as the only variable). Now for some nice easy cases.

## 13.3  TAKING OUT THE MEAN

We have not taken out the general mean but if we choose to be really arbitrary we can say,

'let $x_1 = 1$ for every object'

[remember the $x$'s were any measurements; $x_1$ could well be (number of legs) ÷ 4]. The first of equations 13.1 is now a bit simpler, viz.

$$\Sigma y = b_1(n) + b_2 \Sigma x_2 + b_3 \Sigma x_3$$

this would be much nicer if $\Sigma x_2 = 0 = \Sigma x_3$. Why not? Whatever $x$ is let us throw it away and use instead a measurement $x_2'$ chosen so that

$$x_2' = x_2 - \bar{x}_2 \quad (\bar{x}_2 \text{ denotes the mean of the } x_2\text{'s of all the cows})$$

Similarly for $x_3$.

Then we have (at last)

$$\Sigma y = n b_1 \text{ and so } b_1 = \text{general mean}$$

In case you think what I did in the last paragraph was high-handed here is a bit of detail.

When $x_1 = 1$ for each object (e.g. cow) the equations of estimation (13.1) are

$$\left. \begin{array}{l} \Sigma y = b_1(n) + b_2 \Sigma x_2 + b_3 \Sigma x_3 \\ \Sigma y x_2 = b_1 \Sigma x_2 + b_2 \Sigma x_2^2 + b_3 \Sigma x_2 x_3 \\ \Sigma y x_3 = b_1 \Sigma x_3 + b_2 \Sigma x_2 x_3 + b_3 \Sigma x_3^2 \end{array} \right\} \quad (13.2)$$

From the first of these equations, multiplying by $\bar{x}_2$, we have $\bar{x}_2 \Sigma y = b_1(n\bar{x}_2) + b_2\bar{x}_2 \Sigma x_2 + b_3\bar{x}_2 \Sigma x_3$. Subtract this from the second equation (left hand side from left, right from right):

$$\Sigma yx_2 - \bar{x}_2 \Sigma y = b_1(\Sigma x_2 - n\bar{x}_2) + b_2(\Sigma x_2^2 - \bar{x}_2 \Sigma x_2)$$
$$+ b_3(\Sigma x_2 x_3 - \bar{x}_2 \Sigma x_3) \qquad (13.3)$$

To tidy this up we first need a digression.

Since by definition $\bar{x}_2 = (\Sigma x_2)/n$

$$\Sigma x_2 - n\bar{x}_2 = 0$$

or (subtracting one of the $\bar{x}_2$'s from each of the $x_2$'s)

$$\Sigma(x_2 - \bar{x}_2) = 0$$

Now

$$\Sigma(x_2 - \bar{x}_2)^2 = \Sigma(x_2 - \bar{x}_2)x_2 - \bar{x}_2 \Sigma(x_2 - \bar{x}_2)$$
$$= \Sigma(x_2 - \bar{x}_2)x_2 - \bar{x}_2(0)$$
$$= \Sigma x_2^2 - \bar{x}_2 \Sigma x_2$$

(which appears multiplied by $b_2$ in equation 13.3).

Similarly

$$\Sigma(x_2 - \bar{x}_2)(x_3 - \bar{x}_3) = \Sigma(x_2 - \bar{x}_2)x_3$$

and so we can rewrite equation 13.3 as

$$\Sigma y(x_2 - \bar{x}_2) = b_2 \Sigma(x_2 - \bar{x}_2)^2 + b_3 \Sigma(x_2 - \bar{x}_2)(x_3 - \bar{x}_3)$$

Now

$$\Sigma \bar{y}(x_2 - \bar{x}_2) = \bar{y}\Sigma(x_2 - \bar{x}_2) = 0$$

and so we have, finally

$$\Sigma(y - \bar{y})(x_2 - \bar{x}_2) = b_2 \Sigma(x_2 - \bar{x}_2)^2 + b_3 \Sigma(x_2 - \bar{x}_2)(x_3 - \bar{x}_3)$$

Now if instead of our original sets of numbers $y, x_1(= 1), x_2, x_3$ ($n$ of each) we started with

$$y' \text{ defined as } y - \bar{y}$$
$$x_2' \text{ defined as } x_2 - \bar{x}_2$$
and
$$x_3' \text{ defined as } x_3 - \bar{x}_3$$

(again $n$ of each; we ignore $x_1$ now) and did a multiple regression analysis we should find as equations of estimation exactly the ones we have just produced. So: 'taking out the mean' by using a dummy variate ($x_1$ in the example) with value 1 for every object, is exactly equivalent to discarding the original $y$'s and $x$'s and replacing each one by the deviation from its respective mean value. This latter is, of course, the common way of doing regression calculations (though there are times when it is correct to work with the original $y$'s and $x$'s, *not* their deviations). If you have followed the argument you will I hope have realised something of the equivalence of the two ways of looking at analysis — the regression-mode and the comparison-mode one might perhaps call them.

A more complicated example of this sort of manipulation of regression equations will be given in Section 13.7.

## 13.4  A SIMPLE FIELD EXPERIMENT

Suppose we have done an experiment on six plots with treatments A and B, arranged thus in three blocks of two:

| Block | I | | II | | III | |
|---|---|---|---|---|---|---|
| Plot | 1 | 2 | 3 | 4 | 5 | 6 |
| Treatment | B | A | A | B | B | A |
| Yield | $y_1$ | $y_2$ | $y_3$ | $y_4$ | $y_5$ | $y_6$ |

Why should we not regard the analysis of this experiment as a problem in multiple regression?

We can define variates $x_0, x_1, x_2, x_3, x_4$ as follows:

$x_0 = 1$ for every plot

$x_1 = 1$ for plots in block I, 0 for other plots

$x_2 = 1$ for plots in block II, 0 for other plots

$x_3 = 1$ for plots in block III, 0 for other plots

$x_4 = -\frac{1}{2}$ for plots of treatment A, $= \frac{1}{2}$ for plots of treatment B

(This is more convenient than giving values 0, 1).

There is a snag here. On every plot

$$x_1 + x_2 + x_3 = x_0$$

i.e. these $x$'s are not 'linearly independent' but the next section will show that this does not matter; we can proceed exactly as if they were linearly independent.

To be quite explicit here is the full layout:

| Regression coefficient | Variate | Plot numbers | | | | | |
|:---:|:---:|:---:|:---:|:---:|:---:|:---:|:---:|
| | | 1 | 2 | 3 | 4 | 5 | 6 |
| $b_0$ | $x_0$ | +1 | +1 | +1 | +1 | +1 | +1 |
| $b_1$ | $x_1$ | +1 | +1 | 0 | 0 | 0 | 0 |
| $b_2$ | $x_2$ | 0 | 0 | +1 | +1 | 0 | 0 |
| $b_3$ | $x_3$ | 0 | 0 | 0 | 0 | +1 | +1 |
| $b_4$ | $x_4$ | $+\frac{1}{2}$ | $-\frac{1}{2}$ | $-\frac{1}{2}$ | $+\frac{1}{2}$ | $+\frac{1}{2}$ | $-\frac{1}{2}$ |

The equations reduce to

$$\left.\begin{aligned}
\Sigma yx_0 &= 6b_0 + 2b_1 + 2b_2 + 2b_3 \\
\Sigma yx_1 &= 2b_0 + 2b_1 \\
\Sigma yx_2 &= 2b_0 + 2b_2 \\
\Sigma yx_3 &= 2b_0 \qquad\qquad + 2b_3 \\
\Sigma yx_4 &= \qquad\qquad\qquad (3/2)b_4
\end{aligned}\right\} \qquad (13.4)$$

and of course

$$\Sigma yx_0 = \text{total of all yields}$$
$$\Sigma yx_1 = \text{total of block I}$$
$$\Sigma yx_2 = \text{total of block II}$$
$$\Sigma yx_3 = \text{total of block III}$$
$$\Sigma yx_4 = \tfrac{1}{2}\,(\text{total of B plots} - \text{total of A plots})$$

Note that if we add up the second, third and fourth of equations 13.4 (left-hand side to left-hand side, right-hand side to right-hand side) we get the first equation of the set. This sort of thing always happens when we start off with a 'superfluous' variate (here any one of $x_0$, $x_1$, $x_2$, $x_3$ can be considered superfluous since if we are given the values of any three we know the value of the fourth). 

We now put in the obvious condition

$$b_1 + b_2 + b_3 = 0$$

and we have the usual estimates

$$b_0 \text{ (usually called } m) = (\Sigma y)/6$$
$$b_1 = B_1/2 - m \ (B_1 = \text{total of Block I}) \text{ etc.}$$

and
$$b_4 = (\text{mean of B plots}) - (\text{mean of A plots}).$$

Now the reduction in the total sum of squares ($\Sigma y^2$) is

$$m\Sigma y + b_1 \Sigma yx_1 + b_2 \Sigma yx_2 + b_3 \Sigma yx_3 + b_4 \Sigma yx_4$$

which comes to the same as in Analysis of Variance.
*But* note that in this sort of case (when $\Sigma x_1 x_2 = \Sigma x_1 x_3 = 0 =$ all the
other sums of cross products we could have fitted (for example) $m$ only
and got the same equation

$$m = (\Sigma y)/6$$

and a reduction of $m\Sigma y$.
An extension of this argument shows that we can reduce the original
$\Sigma y^2$ as follows:

(1) fit $m$ only,   reduction $m\Sigma y$
(2) fit $m\ b_1\ b_2\ b_3$   reduction $m\Sigma y + b_1 \Sigma yx_1 + b_2 \Sigma yx_2 + b_3 \Sigma yx_3$
(3) fit $m\ b_1\ b_2\ b_3\ b_4$   reduction $m\Sigma y + b_1 \Sigma yx_1 + b_2 \Sigma yx_2 + b_3 \Sigma yx_3 +$
$+ b_4 \Sigma yx_4$

Or we could do the steps in a different order, say

(1) $m$
(2) $m\ b_4$
(3) $m\ b_4\ b_1\ b_2\ b_3$

I have now shown in some detail (1) the sort of algebraic process that
underlies all regression calculations (in Section 13.2) and (2) how by
using 'dummy' variates this algebra can be made to deal with a designed
experiment; and that, in this case, it leads to an analysis of variance,
treatment means and standard errors exactly the same as those obtained
by the usual sort of analysis in terms of contrasts, effects and so on.
This isn't just a piece of high-sounding theory; it has practical value.
One example is that it gives us general rules for dealing with missing
plots (see Section 13.9). Another is that, if you haven't a computer
program for conventional analysis of variance, but you do have a good
one for multiple regression, you should be able to coax it to analyse
designed experiments reasonably well. Analysis of covariance (see
Chapter 14) is a sort of hybrid of conventional analysis of variance and
regression analysis and can be looked at as a development of either.
The remainder of this chapter will give, with the minimum of
algebra and symbols, hints of some of the applications of regression
analysis in field experiments.

## 13.5 ORTHOGONALITY

The last section illustrates the fact that if several sets of variates are mutually orthogonal — in the example just given the variates fall into three mutually orthogonal sets, namely $(x_0)$, $(x_1, x_2, x_3)$ and $(x_4)$ — then we can legitimately calculate the sum of squares for each set ignoring the others and this sum of squares is unaltered if we take into account at the same time one or more of the other sets (for a simple example of the opposite case see Section 12.6). In other words, if sets of variates are mutually orthogonal the corresponding sums of squares are additive. If not, not. More correctly: if not, in general not.

This is a pretty fundamental thing — the fact that orthogonality is a necessary and sufficient condition for the additivity of sums of squares and I think it is worth a more detailed illustration. Return to the case given early in this chapter — the relation of a 'dependent variate' $(y)$ to three 'independent variates' $(x_1, x_2, x_3)$ about which we assume one thing only — that there is no linear relation between them. The $x$'s now may be observed (uncontrolled) variates (length of cow's tail) or dummy variates (0 for male, 1 for female).

I repeat the 'normal equations' or 'equations of estimation' (13.1) given in Section 13.2:

$$\left. \begin{array}{l} \Sigma yx_1 = b_1 \Sigma x_1^2 + b_2 \Sigma x_1 x_2 + b_3 \Sigma x_1 x_3 \\ \Sigma yx_2 = b_1 \Sigma x_1 x_2 + b_2 \Sigma x_2^2 + b_3 \Sigma x_2 x_3 \\ \Sigma yx_3 = b_1 \Sigma x_1 x_3 + b_2 \Sigma x_2 x_3 + b_3 \Sigma x_3^2 \end{array} \right\} \quad (13.5)$$

First, if $x_1$, $x_2$ and $x_3$ are mutually orthogonal, then

$$\Sigma x_2 x_3 = \Sigma x_1 x_3 = \Sigma x_1 x_2 = 0$$

and the equations become

$$\left. \begin{array}{l} \Sigma yx_1 = b_1 \Sigma x_1^2 \\ \Sigma yx_2 = b_2 \Sigma x_2^2 \\ \Sigma yx_3 = b_3 \Sigma x_3^2 \end{array} \right\} \quad (13.6)$$

These are exactly the equations we get if we calculate the regression of $y$ on $x_1$ ignoring $x_2$ and $x_3$, then the regression of $y$ on $x_2$ ignoring $x_1$ and $x_3$, and so on. It is clear, too, that the reduction in the total sum of squares due to the multiple regression,

$$b_1 \Sigma yx_1 + b_2 \Sigma yx_2 + b_3 \Sigma yx_3 \quad (3 \text{ d.f.})$$

is equal to the sum of the three sums of squares (1 d.f. each) due to fitting the three separate regressions.

This illustrates that orthogonality is a *sufficient* condition for additivity. Now its necessity is a much trickier thing to prove rigorously, and I am not going to prove it. The proof, which is heavy weather algebraically, fails in certain fairly trivial cases; for instance, if for many of the plots (or cows or whatever) $x_1$, $x_2$ and $x_3$ have the same set of three values, leaving three or less distinct sets of values.

Before dealing with other applications of the 'multiple-regression' approach let's clear away the difficulty of 'superfluous variates' or 'redundant constants'.

## 13.6 REDUNDANT CONSTANTS

In the example (the six-plot experiment) dealt with in Section 13.4 there was one superfluous variate ($x_0$, or $x_1$ or $x_2$ or $x_3$, it does not matter which) because there was one (and only one) linear relation between the $x$'s true for all plots, $x_0 = x_1 + x_2 + x_3$. We ignored this redundancy and found that we had a similar relation between the corresponding 'equations of estimation' — the first four equations were consistent (i.e. they did not contradict each other) and in fact any one of the four could be derived from the other three. Finally, we found that the solutions for the corresponding regression coefficients $b_0$, $b_1$, $b_2$, $b_3$ were not unique (we calculate $b_4$ from the last equation and are left with three equations in four unknowns). We put in the arbitrary condition $b_1 + b_2 + b_3 = 0$ in order to get a convenient solution. (We could have put in $b_0 = 0$ or some other arbitrary condition instead.) But arbitrariness has its limits; $3b_0 + b_1 + b_2 + b_3 = 0$ is *not* permissible in this case. See Healy[33], who gives the mathematics of the general case of 'redundant constants'.

## 13.7 ADDING EXTRA INDEPENDENT VARIATES

We have dealt in Section 13.5 with the case in which, after taking out regressions on say 3 variates we wish to consider what would have happened if we had calculated simple regressions for each independent variate ignoring the other two. We now turn to the opposite situation — in which we want to know the effect of taking into account an extra variate that was omitted in the original analysis.

To be explicit, suppose we have 'yields' $y$ and other variates $x_1$, $x_2$ and $x_3$ for $n$ plots. I shall omit the suffix for plot-number throughout this section.

First let us forget $x_3$ and calculate the partial regression coefficients of $y$ on $x_1$ and $x_2$; call them $a_1$ and $a_2$. The equations of estimation are

$$\left. \begin{array}{l} \Sigma yx_1 = a_1 \Sigma x_1^2 + a_2 \Sigma x_1 x_2 \\ \Sigma yx_2 = a_1 \Sigma x_1 x_2 + a_2 \Sigma x_2^2 \end{array} \right\} \qquad (13.7)$$

and the sum of squares of residuals is

$$S = \Sigma y^2 - a_1 \Sigma yx_1 - a_2 \Sigma yx_2$$

with $(n-2)$ degrees of freedom.

Now if we fit regressions on $x_1$, $x_2$ and $x_3$ simultaneously we get a different set of equations (using the symbols $b_1$, $b_2$ and $b_3$ for the coefficients)

$$\left. \begin{array}{l} \Sigma yx_1 = b_1 \Sigma x_1^2 + b_2 \Sigma x_1 x_2 + b_3 \Sigma x_1 x_3 \\ \Sigma yx_2 = b_1 \Sigma x_1 x_2 + b_2 \Sigma x_2^2 + b_3 \Sigma x_2 x_3 \\ \Sigma yx_3 = b_1 \Sigma x_1 x_3 + b_2 \Sigma x_2 x_3 + b_3 \Sigma x_3^2 \end{array} \right\} \qquad (13.8)$$

and the sum of squares of residuals with $(n-3)$ degrees of freedom is

$$S' = \Sigma y^2 - b_1 \Sigma yx_1 - b_2 \Sigma yx_2 - b_3 \Sigma yx_3$$

We wish to look into two relationships between these two different systems. First, how can we conveniently calculate $b_1$, given that $a_1$ and $a_2$ are already known? And secondly, how can we calculate $S - S'$, i.e. the additional amount 'taken out' of the sum of squares of residuals by fitting the regression on $x_3$?

This involves a lot of tiresome algebra which I am *not* going to give you. But here's an outline:

(1) We need an extra idea. Express $x_3$ as a linear function of $x_1$ and $x_2$ plus an 'error' term; in other words calculate the regression equation relating $x_3$ (dependent variate) to $x_1$ and $x_2$ (independent variate). This may be written

$$x_3 = c_1 x_1 + c_2 x_2 + \text{error}$$

So far, we have merely treated $x_3$ exactly as we've treated $y$ — we've taken out the respective regressions on $x_1$ and $x_2$ leaving ourselves with 'error' parts

$$(y - a_1 x_1 - a_2 x_2) \quad \text{and} \quad (x_3 - c_1 x_1 - c_2 x_2)$$

Note that each of these can be regarded as a variate, and each is orthogonal to $x_1$ and to $x_2$. For example

$$\Sigma(y - a_1 x_1 - a_2 x_2)x_1 = \Sigma yx_1 - a_1 \Sigma x_1^2 - a_2 \Sigma x_1 x_2 = 0$$

from equation 13.7.

(2) Next if you care to do the algebra you can show that the result of eliminating $b_1$ and $b_2$ from equations can be expressed thus:

$$\Sigma(y - a_1 x_1 - a_2 x_2)(x_3 - c_1 x_1 - c_2 x_2)$$
$$= b_3 \Sigma(x_3 - c_1 x_1 - c_2 x_2)^2 \tag{13.9}$$

Note the analogy with the equation we should get by ignoring $x_1$ and $x_2$:

$$\Sigma y x_3 = b_3' \Sigma x_3^2$$

Equation 13.9 is of the same form but involves instead of $y$ and $x_3$ their 'error' parts after 'adjusting for' $x_1$ and $x_2$. This leads directly to the method of calculating the regression coefficient from the 'error' line in analysis of covariance (see Chapter 14).

(3) More algebra (mercifully veiled) shows that

$$S - S' = b_3 \Sigma(y - a_1 x_1 - a_2 x_2)(x_3 - c_1 x_1 - c_2 x_2)$$

and *this* is the analogue of the sum of squares taken out by the regression on $x_3$ ignoring $x_1$ and $x_2$, viz.

$$b_3' \Sigma y x_3$$

So: the sum of squares attributable to the regression on $x_3$ (after fitting regressions for $x_1$ and $x_2$) is $b_3$ multiplied by the left-hand side of equation 13.9 from which we calculated $b_3$.

In this argument I have taken a very simple case — only two independent variates in the first instance, and only one additional one. In fact the more general case ($r$ independent variates initially, $s$ additional ones) is not different in principle.

To sum up: when you wish to fit regressions on a second set of variates after fitting regressions on an initial set, do a straightforward calculation, *after* adjusting the $y$'s and the new variates for the first set.

Note that none of the above assumes any orthogonality at all. The more there is the simpler all the calculations, but it isn't necessary.

We now consider some implications in the analysis of experiments.

## 13.8 NON-ORTHOGONAL EXPERIMENTS

First, in any experiment, whether well designed or ill, with yields from all plots or not, you may calculate multiple regressions on dummy variates, or to use an equivalent phrase, fit constants by the least squares method. Indeed all our usual methods of analysis for tidy designs like randomised blocks are just that, although with all the sets of comparisons we want being orthogonal to one another, we go by rule of thumb and tend to forget the underlying process.

## 13.9  MISSING PLOTS

If you fail to record the yield of just one plot it is obvious that not all the value of the experiment is lost. All the treatments except the one applied to the lost plot have mean yields determined with full accuracy — we might consider as a first resort analysing the experiment ignoring all plots of the same treatment as the casualty. This would lead to complications in some cases, e.g. in a latin square the comparison between rows is not orthogonal to those between columns when a plot is omitted. (For analysis of latin squares with a whole row, column or treatment missing see Yates[50].) The procedure discussed below indicates the lines on which we can use all the information provided by the (imperfect) experiment — and the process which emerges finally is simple in most of the common cases.

We now deal with a non-confounded experiment comprising $r$ blocks of $p$ plots each, $rp = n$ plots in all. Suppose we have 'lost' the first plot of of the experiment and that this belongs to block 1 and treatment 1 (this is quite general really — the actual numbers don't matter).

We have a set of $(n - 1)$ yields $y_2, y_3, \ldots, y_n$. Instead of these consider the set of $n$ yields

$$0, y_2, y_3, \ldots, y_n \qquad (13.10)$$

I could write instead of 0 in the first place any number that took my fancy, 99 say or 999; I do not know the yield of the first plot. Now I allow for a factor that has affected plot 1 only, to produce this arbitrarily assigned yield. That is, we set up a dummy variate ($x_1$ say) with values 1, 0, ..., 0. It is clear (I hope) now that different values of the arbitrary yield (0, 99, etc.) will produce different coefficients of regression on $x_1$ but will not otherwise alter the analysis.

We now perform a multiple regression analysis with independent variates $x_1$ and all the ones we should use for blocks and treatments in the normal case when no plot is 'missing'. (This is in fact equivalent to doing an analysis of covariance using as yield, the values given at 13.10 above, with $x_1$ as independent variate, and taking out blocks and treatments in the normal way.)

We are in fact fitting constants thus. For plot 1

$$(0 =) y_1 = m + b_1 + t_1 + a_1(1) + e \ (x_1 = 1 \text{ on this plot})$$

($m$ = mean; $b_1$, etc. = block constants; $t_1$, etc. = treatment constants. $a_1$ = coefficient of regression of $y$ on $x_1$; $e$ = error).

For the remaining plots

$$y = m + b + t + a_1(0) + e$$

the normal form, the $x_1$ regression not being involved.

Our equations of estimation are:

$(m)$  $\Sigma y = n . m + a_1$

$(b_1)$  $B_1 = p . m + p . b_1 + a_1$ ($B_1$ = total of block 1 and so on)

$(b_2)$  $B_2 = p . m + p . b_2$

and so on,

$(t_1)$  $T_1 = r . m + r . t_1 + a_1$ ($T_1$ = total of treatment 1 and so on)

$(t_2)$  $T_2 = r . m + r . t_2$

and so on,

$(a_1)$  $0 = y_1 = m + b_1 + t_1 + a_1$

From these equations we have (writing $G$ for $\Sigma y$)

$$rB_1 + pT_1 - G = rp . m + rp . b_1 + r . a_1 + pr . m + pr . t_1 + p . a_1 - n . m - a_1$$

$$= n . m + nb_1 + n . t_1 - a_1 + r . a_1 + p . a_1$$

$$= n(m + b_1 + t_1) + a_1(r + p - 1)$$

Substitute for $(m + b_1 + t_1)$ from the $(a_1)$ equation and we find

$$rB_1 + pT_1 - G = n(-a_1) + a_1(r + p - 1)$$

$$= a_1(n - r - p - 1)$$

This gives the formula for a missing yield in this simple case; it can be derived easily in other ways.

Note that the equations for $b_2, b_3, \ldots, t_2, t_3 \ldots$ are just as they would have been if the experiment had been complete.

We can calculate standard errors and make tests of significance of treatment differences exactly by use of the methods appropriate in analysis of covariance (see Chapter 14).

An extension of the method given above will cope with experiments in which two or three or more plots are missing, but the computation soon becomes very heavy for an ordinary man on a desk machine. Computers, of course, can cope — if they are told what to do.

# Chapter Fourteen

# ANALYSIS OF COVARIANCE

## 14.1 GENERAL

The analysis of covariance is a very flexible technique, applicable to a
wide range of problems. The arithmetic is always complicated and
much of it is hard to check. But, more important, the interpretation
of the results has many pitfalls for the unwary (see Section 8.10). I shall
deal only with aspects of covariance likely to be useful in interpreting
field experiments. For a general discussion see for example Kendall[37]
and Biometrics[16]. (By the way, plenty of otherwise good textbooks of
statistics give the subject little or no attention.)

## 14.2 PROCEDURE

I shall consider in this section an experiment laid down in randomised
blocks without confounding. And I shall deal with one independent
variate ($x$) only. More complicated cases will be mentioned later.

Suppose, then, that our simple experiment has $p$ treatments
replicated $r$ times, the design being $r$ blocks of $p$ plots each. And
suppose, in the first place that we ignore the variate $x$ and fit constants:

mean:       $m$

blocks:     $a_1 \ldots a_r$, or briefly $a$

treatments:    $t_1 \ldots t_p$, or briefly $t$;

we take each yield to be made up of the appropriate combination of these and an error term ($e$):

$$y = m + a + t + e$$

(this equation represents $n$ equations

$$y_i = m + a_j + t_k + e_i$$

if plot $i$ is in block $j$ and has treatment $k$).

We minimise the sum of squares of the $e_i$ (over all plots) and get equations of estimation:

($m$)   $G = n \cdot m$                         ($G$ = grand total of $y$)

($a_1$)   $B_1 = t \cdot m + t \cdot a_1$ etc.   ($B_1$ = total of $y$ in first block)

($t_1$)   $T_1 = r \cdot m + r \cdot t_1$ etc.   ($T_1$ = total of $y$ in first treatment)

Remember that what we have done can be regarded as calculating the coefficient of regression of $y$ on various 'dummy' variates; for example the regression on $x_0$ ($x_0 = 1$ for every plot) gives the mean $m$.

Note also that the error terms $e_i$ are orthogonal to all these dummy variates. For $e_i = y_i - m - a_j - t_k$ and, for example,

$$\Sigma e \cdot x_0 = \Sigma y x_0 - m \Sigma x_0 - p \Sigma a - r \Sigma t$$
$$= nm - mn - 0 - 0 = 0$$

Now we have seen in Section 13.7 that if we wish to put into the analysis an extra independent variate we in effect adjust this variate for regression on the first set of independent variates and then the process of fitting is straightforward and the amount by which the residual sum of squares is decreased is calculated very simply. So, in this case, let us throw away $x$ and use instead $x'$ where

$$x' = x - m_x - a_x - t_x + \text{error}$$

$m_x$, $a_x$ and $t_x$ being calculated by minimising the sum of squares of the error terms. $x'$ also is orthogonal to the dummy variates corresponding to $m$, $a$, $t$. And so the effect of fitting $b'$, the regression on $x'$ is to add an equation

$$\Sigma y x' = b' \Sigma (x')^2$$

and to decrease the residual sum of squares by

$$b' \Sigma y x', \text{ i.e. by } (\Sigma y x')^2 / \Sigma (x')^2 \quad (1 \text{ degree of freedom})$$

So far, all we have done is to rewrite in different symbols the sort of thing we did in Section 13.7. But now for a few variations.

(1) Clearly the decrease $b'\Sigma yx'$, compared with the mean square for residual error [with $(p-1)(r-1)-1$ degrees of freedom] allows a test of the significance of the regression on $x'$ (or $x$, it comes to the same thing) after the fitting of $m$, $a$, $t$, or (in the jargon) of the regression on $x$ 'eliminating' blocks and treatments. But for a test of significance of treatments eliminating the regression on $x$ we need a new calculation. If we fit $m$, $a$, $b$ (but not $t$) we get the same old equations for $m$ and $a$ but a different equation for $b$ (because instead of $x'$ we are now using a variate, call it $x''$, defined as $x'' = x - m_x - a_x$ + 'error' (this 'error' includes what we previously regarded as $t_x$ as well as the previous error term). So we have a new value for $b$ ($b''$ say), given by

$$\Sigma yx'' = b''\Sigma(x'')^2$$

and a decrease in the sum of squares of $b''\Sigma yx''$.

So the additional sum of squares corresponding to 'treatments' and regression instead of fitting the regression alone is

$$\Sigma tT + b'\Sigma yx' - b''\Sigma yx'' \quad (p-1 \text{ degrees of freedom})$$

As we have seen in Section 13.5 the second and third terms in this expression are not in general equal.

(2) So far we have fitted a common regression 'within' treatments. We can instead fit separate regressions within each of the $p$ treatments. That is, we can postulate that the regression coefficient will vary from treatment to treatment. This corresponds to using, instead of one independent variate $x$, $p$ separate ones $x_1, x_2, \ldots, x_p$ defined thus:

$x_1 = x$ for all plots of treatment 1

$x_1 = 0$ for all other plots

and similarly for $x_2 \ldots x_p$

this process will lead to regression coefficients $b_1 \ldots b_p$ and a decrease in the sum of squares of

$$S_p = b_1 \Sigma yx_1' + b_2 \Sigma yx_2' + \ldots + b_p \Sigma yx_p'$$

($p$ degrees of freedom)

($x_1'$ now means $x_1$ adjusted for blocks only).

The comparison of

$$S_p - b'\Sigma yx'$$

with $(p-1)$ degrees of freedom with the residual mean square gives a test of the significance of the variation of the regression between treatments.

We may of course group the treatments and fit a common regression coefficient for each group; a particular case is a factorial experiment in which we postulate that the regression varies with some of the factors but not with the others.

(3) We may use instead of $x$ a new independent variate $X$ defined thus: $X_1$ = mean of $x$ for all plots of the treatment that is applied to plot 1, and similarly for $X_2, \ldots, X_n$. This leads to a regression coefficient based on the treatment-means of $x$ and $y$ and ignores within-treatment variation entirely. The regression coefficient (and the corresponding sum of squares) can be calculated from the treatment-means above.

The remaining sum of squares for treatments with $(p-2)$ degrees of freedom compared with the residual mean square — that is the residual mean square of $(r-1)(p-1)$ degrees of freedom, *without* allowing for regression, gives a test of the significance of the deviations of the treatment-means from their private regression on $x$. The same process can of course be done for blocks instead of treatments.

We can assemble what we have done in (2) and (3) into one grand analysis, something like this (though I doubt if this would often be worth the trouble).

| | | DF |
|---|---|---|
| Blocks: | regression | 1 |
| | remainder | $r-2$ |
| | | $\overline{r-1}$ |
| Treatments: | regression | 1 |
| | remainder | $p-2$ |
| | | $\overline{p-1}$ |
| Error: | general regression | 1 |
| | variation of regression between blocks | $r-1$ |
| | variation of regression between treatments | $p-1$ |
| | remainder | $(r-2)(p-2)-2$ |
| | | $\overline{(r-1)(p-1)}$ |
| Total | | $\overline{rp-1}$ |

It is possible, if you have any breath left at the end of such an analysis, to test whether for example, the coefficients of regression calculated at 'Error: general regression' and at 'Treatments: regression' differ by more than can be reasonably ascribed to 'error' but this is difficult ground and even the expert guides seem less than unanimous — see Biometrics[16] (first two papers).

## 14.3  EXTENSION TO OTHER CASES

In fact, with a few modifications, the last section can be applied more generally.

(1) For a design in rows and columns you need two sets of constants instead of the one set we used for blocks.

(2) If there is confounding, provided you fit constants for the unconfounded effects and interactions only, this makes no difference to the analysis (but it does affect the presentation and interpretation of the results — see Section 8.4).

(3) If there are two or more independent variates, the procedures given can easily be extended; it will be necessary first to ensure that all independent variates are mutually orthogonal. This sounds a tall order but an example should make the idea clear: if $x_1$ (i.e. $x_{11}, x_{12}, \ldots, x_{1n}$) and $x_2$ (i.e. $x_{21}, \ldots, x_{2n}$) are two independent variates, and if they are not orthogonal, i.e. $x_{11}x_{21} + x_{12}x_{22} + \ldots + x_{1n}x_{2n} = c \neq 0$ or briefly $\Sigma x_1 x_2 \neq 0$, then we throw away $x_2$ and use instead a variate $x_2'$ defined by the $n$ equations ($i = 1, 2, \ldots, n$) $x_{2i}' = x_{2i} - c'x_{ii}$, where $c' = c/(x_{11}^2 + x_{12}^2 + \ldots + x_{1n}^2)$

Now
$$\Sigma x_2' x_1 = \Sigma x_2 x_1 - c'\Sigma x_1^2$$
$$= c - c$$
$$= 0$$

and all is plain sailing.

The case of split-plot designs (or, saying the same thing in different language, the case of designs in which inter-block information is being used) is more difficult and I shall not discuss it.

## 14.4  FERTILITY TRENDS

Analysis of covariance can be used to test whether the yields of an experiment show evidence of a fertility trend. One or two artificial or 'dummy' variates can be constructed to represent 'position in the field'. If all the plots are in one line (and there are no gaps or irregularities in the layout) a single variate $x$ with values 0, 1, 2, 3, . . . for the successive plots starting from one end is all that is needed. A decent computer program can be asked to allow for linear (or quadratic, etc.) regression on $x$ and tell the experimenter whether the error mean square has been lessened by taking out the corresponding sum of squares for one degree of freedom (linear regression) or two or more degrees of freedom (quadratic, etc.).

If the layout was in several rows of plots two variates $x_1$ and $x_2$ will be needed and in the obvious tidy case each can have simple values 0, 1, 2, 3, . . . . More complicated layouts will require less simple sets of values but all that is needed is to calculate the co-ordinates of the centre of each plot relative to axes through the centre of one chosen plot.

Scale does not matter and scales of $x_1$ and $x_2$ need not be the same. The only important figure that depends on the scale chosen is the regression coefficient itself and the interpretation should be clear. The ratio of the regression coefficient to its standard error is unchanged under a change of scale.

Regressions can be calculated separately or simultaneously for $x_1$ and $x_2$. It may be worthwhile to do a routine analysis of covariance for *all* experimental results with linear regressions on $x_1$ (and simultaneously on $x_2$ if it is used) as a simple 'screening' for fertility trends. (This assumes we have a good well-trained computer available.) A slightly tricky point of 'philosophy' or 'ethics' arises here. If you screen many experiments for evidence of trends and use $P = 0.05$ as a criterion of significance you can expect to find one 'significant' trend in (on average) 20 experiments even when there are *no* true trends at all. So, if you adopt the procedure I suggest, you must accept the risk that in an occasional experiment you will allow for a trend that wasn't really there anyway and in doing so present results in a rather less desirable form than if you had never heard the wretched word 'trend' at all. (In these 'rogue' experiments you will be showing standard errors that are biased downwards, and your treatment means will have an additional component of variance due to the spurious adjustments; this mishap in analogous to the simpler one mentioned in Section 8.1.)

A fair working rule is perhaps:

(1) Screen all experiments for trends.
(2) When one is indicated as significant, try to find a reasonable explanation, e.g. soil variation, differential shelter or disease.
(3) If you do find a plausible explanation go ahead and use adjusted means, standard errors, etc.
(4) If not, forget the trend and use the ordinary analysis.

## 14.5 SYSTEMATIC EXPERIMENTS

This leads me to my last use of analysis of covariance (the Cinderella among the ugly sisters of statistics). This is for analysing the results of systematic experiments. Systematic designs have been advocated and used in a variety of circumstances. These are sometimes used as a simpler alternative to randomised designs, e.g. in fertiliser tests on vegetable crops (Cleaver, Greenwood and Wood[18]) or by necessity, for example in the study of the spread of an airborne disease from a defined source (Jenkyn and Bainbridge[36]). I think that anyone who accepts a systematic design should also acknowledge an obligation to investigate the spatial distribution of 'fertility' (i.e. yield under uniform treatment) as far as it can be revealed by his results. He can usually do this by analysis of covariance using one or more dummy variates and considering linear or more complex regressions on them. He may also (or as an alternative) use as a 'covariate' the distance from a fixed point (regardless of direction) — this may be useful if a disease is thought to have spread from a known small source. In such cases covariance should deepen the experimenter's understanding of his own results. We have come near, I think, to the model-maker's back door.

**Chapter Fifteen**

# TRANSFORMATIONS, MODEL-MAKING AND OTHER PASTIMES

## 15.1 TRANSFORMATIONS – GENERAL

Many statisticians have written about transformations; many humble
people have spent distressful hours transforming data. (I include myself
in *both* categories.) On the whole I think more good has been done
than harm. Now that electronic computers pervade all scientific work
(well, all *respectable* scientific work) and transformations can be called
up with a magic wand, I fear the balance may shift from good towards
harm.

I suggest that anyone tempted to indulge in the transformation of
data should think hard and rigorously beforehand whether his data
need transforming at all, and if so, what he desires of the transformed
data. I distinguish two distinct cases – yields of crops, and other data.

## 15.2 TRANSFORMATION OF YIELDS

Yields are sacred; most other observations are not. What I mean is:
other things being equal, it is equally virtuous to increase a yield of
10 cwt per acre to 15 and to increase a yield of 30 cwt to 35. The
extra 5 cwt feeds the same number of mouths in each case; and it puts
the same amount of extra money into each farmer's pocket.

You may be tempted to forget the sacred quality of yields and
'transform' the yields of your plots (e.g. the use of the square-root
transformation for yields of *Robusta* coffee has been advocated by
Butters[17]). If you succumb to temptation you must realise that you
are in effect giving different 'weights' (in the statistical sense) to the
estimates based on the separate blocks (or rows and columns) of your

experiment. If you use the square-root transformation [i.e. analyse $\sqrt{(yield)}$] or the logarithmic transformation [i.e. analyse log (yield)] you are giving more weight to the blocks with smaller mean yields, less weight to those with greater mean yields.

The only unbiased estimate of the effect of a particular treatment over the area occupied by your experiment is the difference between the unweighted means of the appropriate sets of plots. (I assume your experiment is well designed and well conducted.) In so far as the site adequately represents the field it is in, or all the fields of that crop in your county or whatever, the unweighted estimate is the correct (unbiased) estimate you want (or should want). The only argument I can think of to the contrary is: the more progressive farmers are more likely to change their practices in accordance with your results and they probably tend to have crops better than average — so it might be fair to give extra weight to the blocks with bigger yields. This is the opposite of the effect of transformations commonly used (e.g. square root, and logarithmic). And, any way, it would need a bold man to say how weight should be related to yield — avoid this slippery path!

I suppose that if you have an experiment so well-replicated (and on such a variable site) that you can show that on one set of blocks mean yields and error variance are both greater than elsewhere you might reasonably consider using weighted means (plots of each set of blocks having a weight in inverse proportion to the local error variance). But see Section 8.12 for comments on walking the knife-edge between the devil and the deep blue sea in the analogous problem of a series of experiments.

If you find yourself in this situation, I suggest:

(1) you have indulged in extravagant replication;
(2) you have chosen a pretty poor site for your experiment and I hope you will mend your ways.

Anyway, this is a far cry from using a transformation.

So, however you may yearn to normalise the distribution of yield residuals, or to find a nice tidy 'additive model', do not transform yields. Or, if you *must*, present means of raw (untransformed) yields together with results of significance tests done on the transformed data — but isn't life complicated enough already?

## 15.3 TRANSFORMATION OF NON-YIELD OBSERVATIONS

Many of these are measured in pretty arbitrary scales anyway. Do you prefer pH or 'number of litres of soil solution containing one $H^+$ ion'?

Will you take 'conductivity' and pass me its inverse 'specific resistance'? And if the scale does have a semblance of absolute rightness about it, what we really want is a measure of its effect on the crop. Does a bean plant carrying 20 aphids feel the same about the arrival of 10 more as a plant already wilting under 200 (or 2000)?

In these cases there are two crude reasons for considering the use of a transformation.

(1) To make a more useful analysis of the effects of treatments on the observations.
(2) To demonstrate a more illuminating relation between the observations and the corresponding crop-yields.

These two may be compatible and call for the same transformation — or possibly not.

## 15.4 ANALYSIS OF THE OBSERVATIONS

Ideally the transformed values of the observations should have the following properties.

(1) The effects of different factors (including 'blocks' or 'rows' and 'columns') should be additive; that is, their interactions should be indistinguishable from error variance.
(2) The residuals, after taking out 'blocks' (or 'rows' and 'columns') and 'treatments' should all belong to the same distribution. There should, for example, be no tendency for residuals to be greater on plots where the observations are greater ('variance positively correlated with mean').
(3) The distribution of residuals should be 'normal' (i.e. Gaussian), so making valid the usual tests of significance based on tables of '$t$' and '$F$'.

I am grateful to N. E. Gilbert for pointing out that a transformation that has property (1) is likely to have also property (2) because the residuals arise from the combined action of uncontrolled variation from many sources more or less analogous to the factors included in the experiment.

In practice (as far as one can ascertain from experiments with not very many degrees of freedom or error), a transformation that satisfies (1) and (2) often seems to satisfy (3) also, but this cannot be guaranteed.

## 15.5  RELATION BETWEEN OBSERVATIONS AND YIELDS

Here we are looking at (or for) correlations and more especially regressions. In this case, in contrast with the one considered above, we are *not* primarily concerned with the between-plot variation of the transformed observations but with their 'within-plot' or sampling variation. If one plot has a value very different from that expected (allowing for the block and treatment to which it belongs) this does not matter in the least provided that this is a property of the whole plot. For example, if one plot of an experiment has ten times as many insects as all the rest this will help reveal the relation between yield and number of insects; but if insects are counted on only one plant per plot and in one plot the sampled plant carries ten times as many insects as the other plants on the same plot the calculated regression will be far from the truth.

So the properties we seek in this case are:

(1)   The relation between yields and transformed observations should be linear (i.e. can be represented by an equation of the form

$$y = a + bx$$

$y$ being yield and $x$ the transformed observation); or failing linearity, we seek a reasonably simple alternative, e.g. a quadratic relation such as

$$y = a + bx + cx^2$$

(2)   The within-plot variation of the transformed observations, should be 'homoscedastic' − that is, the sampling variation should be the same for small values as for large.

Property (2) is of mainly theoretical interest as we seldom have enough replicate samples within each plot from which to assess the sampling variance. It is desirable for the following reason.

If the independent variate used in calculating a regression coefficient is subject to sampling error the calculated coefficient is on average smaller than the 'true' one; the correction needed for an unbiased estimate depends on the sampling variance of the independent variate, assuming this is constant (see Yates[51]). If it is not constant, that is, in the case of a field experiment, if the within-plot variance of the observations depends on the mean value for the plot, this adjustment is complicated and probably impracticable. So it is better to transform the observations to a scale on which within-plot variance is independent of mean value.

## 15.6  MODEL-MAKING

We are now on the brink of model-making. This is a tempting swimming pool with a safe shallow end, a more dangerous deep end — and a smooth slippery slope leading from the former to the latter!

In interpreting any but the simplest experiments we almost always assume 'the additive model', i.e. that plot deviations are independent of the treatments applied and so are 'additive' with treatment deviations (see Section 11.7 for thoughts about this assumption). In very many factorial experiments three-factor interactions are confounded with blocks and the results are interpreted with the assumption that the confounded interactions are zero — here, as in all model-making, we are putting a constraint on our freedom in interpreting the data. We may be able to check the validity of the model we are using, but in the cases just mentioned this is usually possible only when we have the results of many experiments available for analysis.

Going a little deeper into the water, but keeping our heads safely above the surface, we may use the idea of a response curve. If we do an experiment with four levels of a nutrient (say N) we can fit any reasonable equation involving four unknown constants, for example

$$y = a + bx + cx^2 + dx^3$$

($x$ denotes the amount of N applied). (That is, we can find numbers $a$, $b$, $c$, $d$ such that this equation is exactly true for the mean yield of the plots that receive each level of N; if the levels are 0, 1, 2, 3, the equations are

$$\left.\begin{array}{l} \text{mean of } N_0 \text{ plots} = a \\ \text{mean of } N_1 \text{ plots} = a + b + c + d \\ \text{mean of } N_2 \text{ plots} = a + 2b + 4c + 8d \\ \text{mean of } N_3 \text{ plots} = a + 3b + 9c + 27d \end{array}\right\} \quad (15.1)$$

and these equations determine $a$, $b$, $c$ and $d$ uniquely.)

Another suitable equation is

$$\log y = a + b \log x + c (\log x)^2 + d (\log x)^3$$

this will of course lead to a different set of numbers $a$, $b$, $c$, $d$. Another relationship that has received some attention recently is a little more complicated to write down but easy to illustrate (*Figure 15.1*). This is formed by drawing the straight line joining the points on the graph plotted for $N_0$ and $N_1$ and similarly the line joining the points representing the yields for $N_2$ and $N_3$ and producing these lines to their intersection (assumed to be between $N_1$ and $N_2$).

Whatever response-curve we assume may be useful in several ways. First, it helps us to interpolate and so predict a precise value of $y$ for any value of $x$ not included in the experiment (e.g. $x = 1\frac{1}{2}$ in the case of levels $x = 0, 1, 2, 3$). Secondly it may be useful in summarising the results of several experiments. For example we might find that $a$ varied from site to site but $b$, $c$ and $d$ were constant to within the limits of

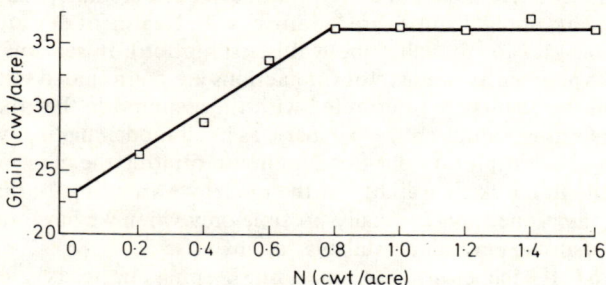

*Figure 15.1* Spring barley: means of 4 experiments

sampling error: or $a$, $b$, $c$ and $d$ might all vary but in proportion. I have just given an example in which the chosen curve fits the data exactly (by 'data' I mean the treatment-means; if we set about 'fitting' a model to the plot yields of all replicates or all blocks things get more compli- cated — and controversial). Whenever you fit a 'reasonable' curve with four unknown constants to four mean yields the fit is exact; and generally a 'reasonable' curve involving $n$ constants can be fitted exactly to any $n$ mean yields. Given $n$ mean yields it is possible to fit a 'reasonable' curve with less than $n$ constants (e.g. you can fit a straight line $y = a + bx$ to any number of mean yields; or of course you can 'fit' a mean value $y = a$ to any number of means — a trivial case). You cannot fit a curve involving more than $n$ constants — if you try you will find that you have too few equations for the constants and the solution is indeterminate. For example, the equations (15.1) above are indeter- minate if any one equation is omitted.

In practice we most often fit a curve with less constants than the maximum the data would allow — we are to some extent simplifying our results in the process. To do this we 'estimate' the constants involved, often by some sort of regression calculation. This may involve trans- formed $x$'s and perhaps a transformed yield ($y$) but otherwise conforms to methods outlined in Chapter 13. There is a more general method known as 'maximum likelihood' (regression analysis is a particular case of 'maximum likelihood') which I shall not discuss — this is the realm of the professional statistician (the deep end of the swimming pool).

The essential point is that in order to make the regression calculation (or to use the method of maximum likelihood) we must assume something about the plot deviations (or at least about the mean of the plot deviations involved in each treatment-mean, i.e. the error component of the treatment-means). The simplest (and commonest) assumption is that all plot deviations are drawn from the same population, perhaps one with a 'normal' distribution. A slightly more complicated (and sometimes more accurate) assumption is that the error variance of a large treatment-mean is larger than that of a small treatment-mean; the variance may be in proportion to the mean value, or perhaps to its square. In other words, we have to assume an 'error structure' and on our choice of error structure depends the values of the constants (or 'parameters') that define the curve (or model) that we are fitting. What with the choice of curve *and* the choice of error structure, we have collected a fair amount of subjectivity by now. Two competent and honest statisticians may well interpret the same experiment very differently. This may be stimulating for statisticians but it is often embarrassing for the humble experimenter.

But those who boldly sport at the deep end of the swimming pool risk slipping into the waste pipe. Some people maintain that a few 'completely-instrumented' experiments can reveal the laws which relate crop yields and the effects of applied treatments to the many uncontrollable (but measurable) conditions of growth. A few such experiments, it is claimed, may give more useful information than a larger number of non-instrumented experiments in which only the effects of treatments on yield are recorded (see Collis-George and Davey[19]).

The argument underlying this approach seems to be that if we know enough about the circumstances of each of a number of similar experiments we can relate the key measurements (say response to a certain rate of N-fertiliser) to such measured 'variates' as temperature, solar radiation and moisture available in the soil. Certainly the yield of a crop, and its response to N, are affected by changes in such 'variates', but what is the formula by which this dependence can be expressed? It is certain that the formula is *not* simple; if the temperature is too low, or too high, yield is zero regardless of other parameters. Some intermediate temperature is optimum (i.e. gives maximum yield) but the optimum temperature with a dry soil may differ from the optimum temperature if the soil is wet. We are dealing with a 'response-surface' which is curved in a complicated way. And what do I mean by 'temperature'? Mean temperature between sowing and harvest? Not very likely, surely, because if season A is $10°$ warmer than B from sowing to mid-season and $10°$ cooler than B thereafter, I for one would not expect the yields to be equal. Right then, let's deal with, say, monthly means, using several separate 'parameters' to describe the 'temperature'

of a season. But why stop at division into months? Potato haulm can
be blackened by one night's sharp frost — the effective leaf area being
set back to zero by a fluctuation that may be balanced by a warm day
later in the month. The fact is that the yield of a crop is a sort of
integral of the conditions in which the crop has grown from moment
to moment of its life and we need an infinity of variates all accurately
measured.

There seems to me two ways out of this slough of despond. First we
can apply deductive reasoning to some set of variates and arrive at a
simple formula giving one new variate that summarises the effects of
the set. An example is the formula (Penman[44]) expressing 'potential
transpiration' as a fairly simple function of temperature, windspeed,
humidity and hours of bright sunshine. Two similar areas of a crop that
both experience the same 'potential transpiration' in a certain period
are believed to fix equal amounts of carbon from the air; at a particular
moment the two crops may experience different temperatures, wind-
speeds and humidities. But, if the calculated 'potential transpiration'
for the period is the same for each crop, then photosynthesis will be
equally effective in both. This, to a professional field experimenter, in
fact to a man whose working life has been devoted to pragmatism, whose
motto might be 'may the sun never set on empiricism' (or even 'suck it
and see'!), is a bit much. He has a cynical feeling in his bones that
biological systems are not as tidily rational at that. (From ignorance,
he may believe that the astronomers and physicists and other 'exact
scientists' live in a different world where constants are constant and
formulae can be guaranteed, but that's another story.) His instinctive
response when faced with a formula purporting to summarise the
effects on a crop of a number of more-or-less independent variables, is
'let's design an experiment to see if it's really true'.

The second possibility seems to be to assume that if any one
parameter ($x$) varies not too much on either side of the average value
encountered in experiments the effect on the crop-yield ($y$) may be
expressed reasonably accurately by some formula such as

$$y = a + bx$$

or
$$y = a + bx + cx^2$$

or perhaps

$$\log y = a + bx + cx^2$$

We are model-making now, not estimating the effects of applied treatments. Putting it another way, $x$ represents an uncontrolled variate
like temperature, not a controlled one like the amount of N applied.

Here $a$, $b$ and $c$ are 'constants', that is they do not vary if all the
other parameters of the crop's environment are held constant. If $x$ =

temperature on 1 May, then $a$, $b$ and $c$ may vary with soil moisture but (over a reasonable range) they do *not* vary with temperature on 1 May.

The use of such simple expressions as I have put on the right-hand sides of the above expressions is justified by the fact that for 'reasonable' curves such formulae are good approximations over a small range (if you want a precise idea of what a 'reasonable' curve is in this context, look up Taylor's theorem in a book of elementary differential calculus). Sometimes it may be necessary to include terms in $x^3$, and perhaps $x^4$ but seldom (in work with crop-yields) any more.

If we generalise the above sort of formula to cope with several variates $x_1$, $x_2$, etc. (i.e. take into account the variation of our 'constants' when other variates than the original one vary) we shall reach a long formula involving powers of each variate and products of powers (e.g. $x_1^2 x_2^2$). A simple case is $y = a + b_1 x_1 + b_2 x_2 + c_{11} x_1^2 + c_{22} x_2^2 + c_{12} x_1 x_2$. (You might like to add terms in $x_1 x_2^2$, $x_1^2 x_2$ and $x_1^2 x_2^2$, but it's complicated enough for me without these.)

If we now decide (pragmatically!) to estimate the constants $a$, $b_1$, $b_2$, etc., from our set of experiments we can use the method of multiple regression (see Chapter 13), treating $x_1$, $x_2$, $x_1^2$, $x_2^2$ and $x_1 x_2$ as five separate 'independent variates'. Adding a dummy variate $x_0$ ($= 1$) to cope with the constant $a$, we have six regression coefficients to estimate. With less than six experiments this is impossible, or more accurately, we can assign an arbitrary value to one or more 'constants' and estimate the rest. You will see, I hope, that unless we have very many experiments we cannot hope to estimate the constants required for even a simple formula involving more than say five variates ($x_1, x_2, \ldots$) if we include their squares and products in our formula. So, to me, this road seems (in the sort of cases I know) to lead inevitably back into the slough of despond. So I conclude with a digression that may cheer you a little.

## 15.7 DANGERS OF MULTIPLE REGRESSIONS

First, as I've hinted above, suppose you have $n$ observations — they may be yields, counts, phone numbers or what you will. Write them down in a row. Below them write down a row of $n$ numbers — they can be anything you like, more phone numbers, your weight in pounds on $n$ successive days, or just a row of 1's. Write down another row of numbers and another row and so on until, below the original observations you have $n$ rows of $n$ numbers. There is one restriction on this — there must be no linear relation between the $n$ rows of arbitrary numbers. Then you will find that it is possible to calculate a formula

$$y = a_1 x_1 + a_2 x_2 + \ldots + a_n x_n$$

($a_1, a_2, \ldots, a_n$ being constants)

which expresses each observation in terms of the arbitrary sets of numbers $x_1 \ldots x_n$ and this formula is exact. This is the bleak side of regression analysis — you can reduce the sum of squares of the differences of a set of observations by taking out a regression on a variate that has *no* relation, causal or otherwise, and the more regressions you take out, the greater the reduction. If you take out enough regressions, the sum of squares remaining will be small — or zero. This fact, I suspect, lurks behind some graphs on which 'actual' and 'predicted' seem to fit so satisfyingly. The question to ask is, how many degrees of freedom were left after fitting the regressions?

## 15.8  COMMON SENSE ABOUT CORRELATIONS

Secondly, there are cases where the correlation of two variates is so great that most people are content to accept it as fact without the paraphernalia of tests of significance. Few people, I think, would demand replication before accepting the reality of the correlation between beheading and death (here we have two variates each of which has only two possible values; we might put

$$y = 0 \text{ (alive)}$$
$$y = 1 \text{ (dead)}$$
$$x = 0 \text{ (head on)}$$
$$x = 1 \text{ (head off)}$$

the correlation isn't necessarily +1 by the way because some of our sample may die with their heads on). This leads me to an analogous case where you may allow yourself to be convinced without a test of significance, if you'll permit a descent into the bathos of common sense. If you apply some treatment to part of a field (a single plot, or a strip, or one end of the field) and a marked difference of growth ensues which corresponds exactly to the boundary of the treated area (and no other treatment has been applied to exactly the same area) you may fairly assume the treatment has caused the effect. I am thinking of a soil very deficient in, say, copper, where a spray of copper can increase the yield of wheat from 2 cwt to 20 cwt per acre (or of the application of a weedkiller that kills a crop). We are nearly back full circle at the 'accidental experiments' I mentioned in Chapter 1.

## 15.9  'COMPLETELY-INSTRUMENTED' EXPERIMENTS

An extreme case of the 'completely-instrumented' experiment is work

done on a single replicate with as few as two (or perhaps one) treatment, with an attempt to measure all important parameters of the
environment of the crop, for example the 'macroplots' described by
French, Long and Penman[29].

Such experiments ('studies' might be a better word) seem to me to
be perfectly valid for some model-making purposes. It may be of great
importance to know that during a week when $x$ calories of solar radiation
were received by a crop whose leaf area index was $y$ its dry matter
content increased by $z$ cwt per acre. Or that in adjacent areas with and
without irrigation the difference in leaf area index was or was not
closely correlated with the difference in dry matter production throughout the season. In such cases the (unassessed) variability from plot to
plot within the experimental field may perhaps be safely ignored. A
portrait of one man does not allow us to compare him with his
relations but it shows detail of his face that we cannot see in a picture
of a family group. The portrait tells us more about the facial
characteristics of men than the group picture even though it presents
a sample of one only. If we want to know whether men have one eye
each or two the portrait is what we need. If we want to know whether
wheat yield is limited by leaf area, by efficiency of the leaf or by the
number of flower-initials formed by the plant a 'two-plot' experiment
with full measurements may be the most useful thing we can do. But
this, to me, is a problem in analysis of time series, not of comparative
field experiments, and so has no place in this book.

# Appendix A

# ROUNDING OFF

Consider an experiment with a mean yield of 100 (in some arbitrary units). Then the likely standard error per plot may be (in my experience) anything between 2 and 50. I will take as an example a standard error per plot of 10, corresponding to an error variance of 100. If we weigh to the nearest 1 unit (instead of weighing exactly) we shall increase the error variance (on average) by 0.083 (= 1/12). (This is an example of the use of Sheppard's correction, which is explained in most statistical textbooks.) This lessens the accuracy of our experimental results in the ratio 100: 100.083. The standard error per plot is 10.0042 instead of 10.0000. Another way of looking at this is that we have lost as much information as if we had discarded 1 replicate in 1201 — hardly a catastrophic loss.

The saving of time in weighing to a relatively coarse degree of accuracy may be negligible but if it saves even one gross error (e.g.

Table A.1 EFFECTS OF ROUNDING OFF PLOT WEIGHTS

| S.E. (%) without rounding | S.E. (%) with rounding (and fraction of information lost) | | |
|---|---|---|---|
| | Rounding to nearest | | |
| | 1% | 2% | 5% |
| 1 | 1.041(1/13) | 1.15(1/3) | 1.76(25/37) |
| 2 | 2.021(1/49) | 2.082(1/13) | 2.47(25/73) |
| 5 | 5.0083(1/301) | 5.033(1/76) | 5.20(1/13) |
| 10 | 10.0042(1/1201) | 10.017(1/301) | 10.10(1/49) |

writing 98.73 for 108.73) it has been well worth while. (See Appendix B for some calculations of the effects of gross errors.)

*Table A.1* shows the effects of rounding in a few cases. Most of the situations commonly encountered are within the range of the table.

Now the standard error per plot is itself subject to sampling variations (that is, if we regard the plot-residuals of an experiment as a sample from a certain population, several experiments sampling the same population will give different standard errors per plot). The standard error of the standard error per plot depends on the standard deviation of the population of residuals ($\sigma$, say) and the number of degrees of freedom for error ($N$) the formula being

$$\text{SE of (SE per plot)} = \sigma/\sqrt{(2N)}$$

*Table A.2* shows some relevant values.

**Table A.2**  STANDARD ERROR OF STANDARD ERROR PER PLOT

| Standard deviation (%) | S.E. of (S.E. per plot) Degrees of freedom | | |
|---|---|---|---|
| | 10 | 20 | 40 |
| 1 | 0.22 | 0.16 | 0.11 |
| 2 | 0.44 | 0.32 | 0.22 |
| 5 | 1.1 | 0.79 | 0.56 |
| 10 | 2.2 | 1.6 | 1.1 |

Reading *Tables A.1* and *A.2* together we see that rounding off the yields affects the standard error per plot by an amount small relative to its sampling variation except where the unit for rounding is greater than the standard deviation (estimated by the standard error per plot).

To sum up: plot-weights can safely be rounded to the nearest 0.5% if the standard error per plot is expected to be 1% or greater, to the nearest 1% if it is expected to be 2% or greater, and so on. And if we get a standard error per plot half as big as expected no great harm will be done by rounding as suggested.

# Appendix B

## GROSS ERRORS

Suppose an experiment contains $r$ replicates of $t$ treatments and that each replicate is laid down as one randomised block. The yields may be written as

$$y_1, y_2, \ldots, y_n (n = rt)$$

I shall use $\bar{y}$ to denote the mean of $y_1 \ldots y_n$.

Now suppose one gross error is made in recording the yields. It seems reasonable to suppose it is equally likely to happen on plot 1 or plot 2 — in fact on any of the $n$ plots. Suppose the error results in one yield being recorded as too much by a number $G$ (positive or negative). Now, averaging over the $n$ possible situations, the effect of the gross error is to increase the sum of squares for any one degree of freedom in the analysis of variance (including the correction for the mean) by $G^2/n$. If you find this brief statement unconvincing, read the next paragraph.

The yields recorded are either

$$y_1 + G, y_2, y_3, \ldots, y_n$$

or

$$y_1, y_2 + G, y_3, \ldots, y_n$$

or any of the remaining $(n - 2)$ possible arrangements. In all cases the mean is $\bar{y} + G/n$. Now any comparison between the $y$'s may be written (when no gross error occurs) as

$$C = l_1 y_1 + l_2 y_2 + \ldots + l_n y_n$$

where

$$l_1 + l_2 + \ldots + l_n = 0$$

165

The sum of the squares derived from the comparison $C$ is

$$S = (l_1 y_1 + l_2 y_2 + l_n y_n)^2/(\Sigma l^2)$$

which can be written $C^2/(\Sigma l^2)$.

If $C'$ is the value of $C$ when the error $G$ is added to $y_1$,

$$C' = C + l_1 G$$

and the sum of squares is now

$$S' = (C + l_1 G)^2/(\Sigma l^2)$$
$$= C^2/(\Sigma l^2) + 2 l_1 GC/(\Sigma l^2) + l_1^2 G^2/(\Sigma l^2)$$

The average value of $S' - S$ (when $G$ occurs equally often on every plot) is

$$[(l_1^2 + l_2^2 + \ldots + l_n^2)/n] G^2/\Sigma l^2 = G^2/n$$

The calculation just done gives us an idea of the average disturbance to a set of results due to the presence of one gross error; it enables us to compare this with other 'disturbances', e.g. those due to coarse rounding off, or to losing a replicate. But remember that a gross error is a sample from a population of an especially nasty nature. Almost any other imaginable source of disturbance (whose average effect is equal to that of a gross error) is preferable.

# Appendix C

# DIRECT-RECORDING BALANCES

'Direct-recording balances' are now becoming available; they produce a record of weights in a form suitable for putting into an electronic computer directly, in one step. The record of the weights from an experiment may be a length of punched paper tape, or a pack of punched cards, probably one for each plot or sub-plot. Such a balance, provided it works well, eliminates the job of reading a pointer and writing down (or typing) the figures read. The balance-record for each plot must include a means of identifying the plot; if this is lacking or wrong the record is almost certainly worthless. The plot can be identified by a number or the convention can be observed that the weights are in the same order as the plot numbers (which in this case need not be recorded).

If you are planning to use a direct-recording balance it is wise to arrange an immediate (or nearly-immediate) print-out from the tape or cards. This should be scanned for obvious errors (or omissions) *before* the produce of the individual plots is disposed of. A dubious figure can then be verified before it is too late. Parallel outputs from the balance (one to the punch, one to a printer) are *not* satisfactory because if the punch misbehaves (e.g. punches 6 instead of 5) this will not be apparent in the printed record.

A well-designed installation (and a suitable editing program available at the computer) can deal with various difficulties.

(1) Missing values (e.g. where the produce of one or more plots has been lost) can be indicated by a special symbol.

(2) A record containing a wrong code number or a wrong weight, if detected immediately, can be cancelled by putting another special symbol into the tape.

(3) A record containing a wrong weight can be superseded by a correct record (with the same plot number) at any stage of the recording of an experiment.

# Appendix D

## STATISTICAL CALCULATIONS – SOME PRACTICAL CONSIDERATIONS

### D.1 ROUNDING OF PLOT-YIELDS, MEANS, ETC.

We have seen in Appendix A that there is usually no gain (but there may be a loss) from recording yields and other data from the plots of an experiment to a finer degree of accuracy than 1 per cent. In case you have to analyse someone else's data which have been recorded to an unnecessary number of 'significant figures' (an unfortunate phrase in this connection) how do you round off?

The crudest technique is simply to delete the unwanted digits; this is easy and foolproof but leads to a slight bias (the yields are diminished by an average of about 5 units in the first discarded place). The best rule seems to be round to the nearest unit up or down and to solve the difficulty of the 5's by rounding to the nearest *even* number, up or down, e.g. 6.5 is rounded to 6, 7.5 to 8. This avoids all bias. This rule should be used always, e.g. in writing down mean yields, plot-factors, etc.

### D.2 THE NUMBER OF FIGURES TO KEEP IN SUMS OF SQUARES, ETC.

In the computation of sums of squares the intrinsic accuracy is that of the square of one plot yield. For example if plot yields are recorded to the nearest unit, sums of squares should be recorded to the nearest unit; division of the square of a total by the number of plots comprising it will lead to decimals which may as well be discarded straight away. If yields have one place of decimals, sums of squares should have two: and so on. When mean squares are calculated of course it is reasonable to introduce a few additional decimal places, one for any divisor up to 10, two for divisors from 11 to 100 and so on.

It is worth while to calculate a mean to one decimal more than is justified by the rounding of the corresponding total; later you may want to calculate means of means or you may want to multiply the mean by a plot factor. In general, keep a 'luxury' decimal throughout in calculating means, etc., and round the final figure at the end of the analysis.

## D.3  CHECKING — OR COMMON SENSE ADNAUSEAM

Assuming that you do your computations on a desk calculating machine (hand or electric) errors can arise in the following ways:

(1)  In reading a number from the data.
(2)  In entering it in the machine.
(3)  In the working of the machine.
(4)  In reading a number from the machine.
(5)  In writing it down.

Errors in the function of calculating machines are rare; when they do occur they are sometimes (but not always) absurdly obvious — such as the unexpected appearance of a string of 9's or the complete jamming of a register. Subtle errors, such as a failure to carry, are another matter.

To repeat a computation *ab initio* and compare the final results with those of the original work is an almost complete check. If, however, the two calculations are done by the same person by the same steps he may make the same error twice, e.g. in reading a number from the original data, or in a faulty 'short-cutting' trick on the machine.

If you cannot arrange for the duplication of all your computations then you should check a few of the most important parts of an analysis, e.g. the correction for the mean, and you should arrange as often as possible that one computation is checked by another, e.g. if you have calculated the total of your observations and later you compute the sum of their squares you can on some machines accumulate the sum of the observations on the second register of your machine. On many electric machines squaring is automatic and each observation has to be entered once only to produce both the sum of squares and the total — a useful feature. On simpler electric machines and all hand machines a number has to be entered twice in order to produce its square. An error can occur in either entry.  If it occurs in the multiplier it will produce an error in the total as well as in the sum of squares (on the multiplier register, which I assume is *not* cleared between operations), if in the multiplicand it will give an error in the sum of squares only.  With this latter class of machines the verification of the total does *not* completely check the calculation, with the former it does.

Here is an example of a computing scheme which is economical in time and checks itself to a reasonable extent; it is for an experiment in randomised blocks.

(1)  Calculate the sum of the squares of the observations ($O$); take them in block order and clear the second register only at the end of each block, first writing down each block total.

(2)  Calculate the sum of the squares of the block totals. Read off the grand total from the second register, clear the register and divide the sum of the squares by the number of plots per block. Write down the quotient ($B$).

(3)  Calculate and write down the various treatment totals.

(4)  Sum their squares and verify the grand total from the second register. Divide the sum of the squares by the number of replications and write down ($T$).

(5)  Calculate the correction for the mean (CFM) and enter it negatively in the product register.

(6)  Add in $B$, write down the sum of the squares for blocks, subtract $B$ (which has remained on the keyboard or drum) to restore the negative CFM.

(7)  Add in $T$ and write down the sum of squares for treatments. Subtract $T$.

(8)  Add in $O$ and obtain the 'total sum of the squares'.

(9)  Subtract the sums of the squares for blocks and treatments to obtain the residual sum of the squares.

(10)  Calculate the mean square for error and the standard error per plot and so on.

If you have a very steady nerve and a tidy mind you may like to modify the scheme just given, by working the whole calculation with every sum of squares $n$ times as great ($n$ being the number of plots). The results of the steps (with the usual symbols) are then as follows.

(1)  $n\Sigma y^2$

(2)  $\Sigma B^2$ (no division needed)

(3)  $T_1$ . . . (as before)

(4)  $\Sigma T^2$ (no division needed)

(5)  $-(\Sigma y)^2$ (no division needed)

(6)  $n \times SS$ (blocks)
(7)  $n \times SS$ (treatments)
(8)  $n \times SS$ (total)
(9)  $n \times SS$ (error)
(10) Divide by $(n - 1)$ so get $n \times MS$ (error) − if you wish to make $F$-tests.
(11) Divide by $n$ for $MS$ (error) in usual units.
     (Steps 10 and 11 can be done in one if $F$-tests are not required.)

When you calculate treatment means in the final units you should calculate the marginal means from the marginal totals (with an appropriate factor). Then check that each marginal mean agrees with the figures it refers to and the whole table is checked.

## D.4  DEVICES FOR EASING THE BURDEN

(1)  Use of a working mean. You can subtract any convenient number from all the yields of an experiment before analysis. You can forget about this until you calculate treatment means for your final tables, and the SE (%).
(2)  Use of a table of squares (and, for that matter, one of square roots).
(3)  Calculating squares and square roots by slide-rule (also very useful when applying a constant plot-factor to a number of means).
(4)  Calculating square roots by the 'odd number method'. This is very old and well-known but I have never seen it given in a book so I will describe it here. [It works on a machine with a keyboard or a drum (e.g. Brunsviga) in which you can modify the 'multiplicand' one figure at a time without clearing the whole register.]

Enter the number $(n)$ whose square root you want in the product register. Pair off the figures starting at the decimal point − as in calculating the square root on paper.

Enter 1 in the multiplicand register below either the *first or second* digit of $n$, whichever is the right-hand of a pair. Subtract, alter the 1 to 3, subtract, alter the 3 to 5, subtract, and so on until you see that subtracting the next odd number would give a negative number in the product register. (You may need to go beyond 9, e.g. subtract successively 1, 3, 5, 7, 9, 11, 13.) Now add 1 to the number standing in the multiplicand register making it an even number (in fact it will equal twice the number of subtractions you have done so far). If you overshoot and accidentally get a negative number by doing one too many subtractions it is easy enough to add back in the number you subtracted in error, and then *decrease* the 'multiplicand' by 1. Move the product register one place to the left (or the multiplicand one place to the right, depending on the machine). Enter a 1 in the place next to the right of where you started (do *not* alter the figure(s) already set). Subtract. Change the 1 to 3, subtract. And so on. The square root of $n$ appears, figure by figure, in the multiplier (or quotient) register. Apart from perhaps needing rounding, it is as accurate as it appears each time you are ready to shift the register.

This method depends on the fact that the sum of $n$ terms of the series of odd numbers 1, 3, 5 . . . ($2n$ + 1) equals $n^2$.

Here is an example: $A$, $B$ and $C$ represent the registers as they appear in many machines, and we are calculating $\sqrt{2}$.

$A$ (top)      = quotient or multiplier.
$B$ (middle) = dividend or product.
$C$ (lower)  = divisor or multiplicand.

| | A | 1 | 4 | 1 | 4 |
|---|---|---|---|---|---|
| | B | 02 | 00 | 00 | 00 |
| | C | 01 | 00 | 00 | 00 |
| | B | 01 | 00 | 00 | 00 |
| | C | 00 | 21 | 00 | 00 |
| | B | 00 | 79 | 00 | 00 |
| | C | 00 | 23 | 00 | 00 |
| | B | 00 | 56 | 00 | 00 |
| | C | 00 | 25 | 00 | 00 |
| | B | 00 | 31 | 00 | 00 |
| | C | 00 | 27 | 00 | 00 |
| | B | 00 | 04 | 00 | 00 |
| | C | 00 | 02 | 81 | 00 |
| | B | 00 | 01 | 19 | 00 |
| | C | 00 | 00 | 08 | 21 |
| | B | 00 | 00 | 90 | 79 |
| | C | 00 | 00 | 28 | 23 |
| | B | 00 | 00 | 62 | 56 |
| | C | 00 | 00 | 28 | 25 |
| | B | 00 | 00 | 34 | 31 |
| | C | 00 | 00 | 28 | 27 |
| | B | 00 | 00 | 06 | 04 |

and so on.

## Appendix E

## THE INTELLIGENT CUSTOMER'S GUIDE, OR WHAT TO ASK YOUR COMPUTER

If you are served by an electronic computer you can reasonably expect more detail or thoroughness in the routine analysis of experiments than you would accept if you yourself were doing the analysis with, say, an abacus.

Here are some suggestions for the things you may want in the analysis.

(1) A clear statement of how the raw data (yields in original units, determinations of dry matter, dirt-tare, ginning out-turn or whatever is appropriate) have been modified and combined to give the figures that go into the analysis proper. A statement of the maximum, minimum and mean values of each variate (original or derived) is useful.

(2) An intelligible 'plan' of the experiment showing the treatments in the order in which the plot-data were taken in.

(3) A table of the yields that have been analysed (e.g. grain at 85% dry matter in kg/ha) in standard order. This is useful if you want to do some extra-curricular calculations, or wish to examine the yield of a particular plot which your visual scores suggest might be anomalous.

(4) A statement of the factors included in the analysis, of any factors that have been ignored and of any dummy comparisons that have been taken into account.

(5) A statement of any plots treated as 'missing' and of the yields assigned to them.

(6) A tabular statement of the plot-residuals; a glance at this, especially if it can be arranged in rows and columns as on the ground, will sometimes reveal a fertility trend which can later be allowed for. Isolated large residuals can occasionally lead to the detection of gross errors.

(7) Full-dress analysis of covariance when 'independent variates' are available that may help to reveal how the treatments produced their effects of yield.

## Appendix F

## GLOSSARY

AFTERMATH Commonly used of the grass that grows during the later part of the season after a cut of hay or silage has been taken. Although the word means 'after-mowing' the phrase 'aftermath grazing' is normal parlance.

ALFALFA (U.S.A. and Canada) = lucerne (*Medicago sativa*) (U.K.)

ALLEY (U.S.A.) = PATH (U.K.)

ARABLE Literally, ploughable. Arable land is land which is already ploughed ('tillage') or is likely to be ploughed in the fairly near future. ('Ploughed' here includes 'treated chemically to allow the sowing of a tillage crop'.) We may state an equation:

$$\text{Arable} = \text{tillage} + \text{leys}$$

Non-arable farm land includes permanent grass, orchards, woodland, rough grazing (e.g. hills), roads and tracks.

AREA I take the area of a plot of a crop grown in rows to be the product length of plot x distance between rows x number of rows and so includes a half-space beyond each edge-row. Similarly for the harvested area, usually with a smaller number of rows. To save putting down two strings instead of one, the manured area of a plot sometimes includes half the path-width each side.

ARRANGEMENT See LAYOUT.

'BALANCING MANURING' In some cases when a fertiliser is tested on sub-plots (e.g.-*v*. P) of a crop in a rotation the RESIDUAL effect is destroyed by applying 'balancing' manure (e.g. P *v*. −) to the following crop of the rotation. A pious hope?

BARNYARD MANURE (BYM) (U.S.A.) = FARMYARD MANURE (FYM) (U.K.)

BASAL Distinguishes an operation which is applied to all plots of an experiment − basal manure, basal spray, etc. In a rotation experiment where all the plots in one SERIES are in the same crop the word basal may reasonably be used of any operation applied to a series; in ROTATION or CROP-SEQUENCE experiments where different treatment-crops occur together it seems better to speak of the 'standard' operations on any one crop.

Unfortunately many horticulturists use 'base fertiliser' to mean fertiliser applied before sowing or planting, as distinct from top-dressings applied later. Perhaps it's a good thing I've never been asked to do an experiment on tomatoes.

BASIN An area of land surrounded by a bank, BUND or border (U.S.A.) that is flooded during irrigation (usually by flow from a ditch or 'watercourse'). The soil surface in a basin must be level to within about 2 in. (5 cm) or less; a typical application of water is equivalent to about 4 in. (10 cm) depth. (These remarks do *not* apply to 'floating' 'floating' varieties of rice.)

BLOCK (often used for 'randomised block' but I suppose blocks could exist in systematic designs, and some blocks have 'restricted' RANDOMISATIONS) A set of PLOTS, to which a certain set of TREATMENTS is assigned. Usually a block is compact in LAYOUT on the field (since plots close together may be expected to be more alike than plots further apart) but this is not essential.

BORDER-EFFECT (U.S.A.) = EDGE-EFFECT (U.K.)

BUND (U.S.A. border) The bank round an irrigation BASIN; the word is also used for the basin. Bunds are often about 2 ft wide and 2 ft high. Walking along a bund between two full basins is exhilarating but beware soft spots.

CHECK (PLOT) (U.S.A.) = CONTROL (PLOT) (U.K.)

COEFFICIENT OF VARIATION See STANDARD ERROR.

COMPETITION EFFECT In a field crop each plant competes with its neighbours for light, water, nutrients, etc. If two adjoining plots have treatments that cause different types of growth, the edge-row of the plot with the more vigorous crop will gain at the expense of the weaker row next to it and so the difference in yield recorded (if no DISCARDS are allowed at harvest) will over-estimate the effect of the treatment, assuming that the more vigorous crop gives the better yield.

CONFOUNDED, CONFOUNDING  A FACTOR or INTERACTION is said to be confounded with BLOCKS if the TREATMENTS are grouped together in the blocks exactly as they are grouped to calculate the EFFECT or interaction concerned. A factor or interaction is partially confounded if it is confounded in some but not all REPLICATES of an experiment.

Where one or more of the factors involved has three or more levels there is another complication. For example, there are four DEGREES OF FREEDOM in the interaction between two factors each at three levels and these are associated in pairs. If we wish to arrange the nine treatments in three blocks of three without confounding the main effect of either factor we can confound either one pair or the other. The two arrangements that result are (taking the factors as $A$ and $B$):

| | | |
|---|---|---|
| Block I | $a_0 b_0, a_1 b_1, a_2 b_2$ | Pair confounded |
| II | $a_0 b_1, a_1 b_2, a_2 b_0$ | $AB(I)$, also called $AB^2$ |
| III | $a_0 b_2, a_1 b_0, a_2 b_1$ | |

and

| | | |
|---|---|---|
| I | $a_0 b_0, a_1 b_2, a_2 b_1$ | |
| II | $a_0 b_1, a_1 b_0, a_2 b_2$ | $AB(J)$, also called $AB$ |
| III | $a_0 b_2, a_1 b_1, a_2 b_0$ | |

In either arrangement the interaction $AB$ may be said to be partially confounded.

CONTROL PLOT, CONTROL  Potentially confusing words (insecticides may control the pest on all except the control plots!) whose use should be reduced to a minimum. Use 'untreated' or 'nil plot', or, as in U.S.A., 'check plot'.

CORN  (In U.S.A. = maize) In U.K., used loosely of any grain crop (as in 'up corn, down horn', i.e. more crops, less livestock); mainly of wheat, oats, barley and rye. 'Mixed corn' = oats and beans (*Vicia faba*) and other mixtures, known locally as 'mushlum'. 'Barleycorn' = a single grain of barley; 'thousand corn weight' is a similar usage.

'CORRECTIVE' MANURING  In ROTATION EXPERIMENTS comparing different rotations an unbalance in mineral nutrient status may develop between rotations. This can be evened out by suitable corrective manuring applied differentially to bring all the soils of the rotations to the same mineral nutrient status. The rates of application of corrective dressings may be based on soil analyses or on analysis of the produce removed in previous crops.

COURSE  Of crop-rotations. 'Three-course', four-course', etc., are clear enough if there is exactly one crop per year. (Fallow for a full cropping season is eligible as a 'course'.) But in tricky cases (e.g. a horticultural experiment may have three crops in two years) avoid 'course'.

CROP-SEQUENCE EXPERIMENT  An experiment lasting two or more years, not designed to continue indefinitely (contrast ROTATION experiment). In any one year the same or different crops may be grown on the several plots.

CUMULATIVE  Of a treatment applied to several successive crops on the same plot(s). Hence 'cumulative effect'.

CYCLES  In ROTATION EXPERIMENTS, used to denote one rotation ('first cycle', 'later cycles'). In some experiments treatments and crops run in cycles of different lengths.

DEGREE OF FREEDOM  This odd phrase derives from a geometric way of thinking about statistical analysis (see Chapter 12); a point that is confined to a plane has two degrees of freedom ('2 d.f.'), one confined to a line has 1 d.f. The yields from an experiment on (say) 32 plots can be imagined to be any 32 non-negative numbers (in a sense there are 32 degrees of freedom). For most purposes we are interested in differences of yields between plots or sets of plots, not in the value of the mean yield of all plots. If then we tabulate the 32 deviations from the mean we have lost none of the information we want, but now the sum of the 32 figures must be exactly zero. If we write down the first 31 out of our imagination (positive, negative or zero in this case) then the 32nd one is settled for us – we have 'lost a degree of freedom'. What we have done corresponds exactly to deducting the 'correction for the mean' (1 d.f.) from the sum of squares of the yields in the analysis of variance.

DESIGN  Given the TREATMENTS to be included in an experiment, and the number of REPLICATIONS, we do not usually just randomise the whole lot. The restrictions (e.g. separation of replicates, arrangement in BLOCKS, CONFOUNDING, LATIN SQUARES) imposed by the guardian statistician or well-tutored experimenter are the design. This is not to say that the statistician may not have forceful (and useful) things to say about the choice of treatments, or about the desirable number of replications. See also RANDOMI-SATION.

DIRECT (effect) The effect of a FACTOR in the YEAR of application is said to be direct; contrast RESIDUAL effect.

DISCARD The area within a plot that receives the appropriate treatment but is not harvested for yield. Discard thus includes GUARD-ROWS, but may include more, e.g. if the machine used for harvest cuts a strip much narrower than the treated plot width.

DUMMY COMPARISON In a factorial experiment the zero level of one factor may render another factor meaningless. For example, if an experiment on rates and times of application of N is designed as a 3 x 3:

Rates:   0, 1, 2 cwt N/acre
Times:   seedbed (S), top-dressed early (E), top-dressed late (L)

there will be three plots per replicate without N and these three are treated identically. So the comparisons OS v. OE v. OL (2 DEGREES OF FREEDOM) are 'dummy'. Note that the same treatments could also be dealt with as:

Rates:   1, 2 cwt N/acre
Times:   never (0), seedbed (S), early (E), late (L).

This gives eight plots per replicate with one dummy comparison. See Section 3.1.

DUNG Often used as a synonym of FARMYARD MANURE but I think it is better used in the sense of 'faeces'.

EC Electrical conductivity. A measure of the salinity of soils (and waters). The EC of a soil is measured on the 'saturation extract' or on mixtures of soil and water of specified ratios. The EC of a soil differs according to which method is used. EC is the inverse of 'electrical resistivity', which is the resistance between two electrodes each 1 cm square, 1 cm apart, immersed in the extract at temperature 25°C. ECs of soil extracts are expressed as 'millimho per cm' (mmho per cm, or mmho for short) but no doubt the adoption of SI units will put a stop to *that*.

1 mmho per cm = 0.001 mho/per cm, i.e. the reciprocal of 1000 ohm cm. Most crops grow well in soil with EC less than 4, badly with EC greater than 8.

A logarithmic scale, using $pC = -\log_{10}$ (EC in mho per cm), is also used. I suspect that in replicate samples from areas of differing salinity the VARIANCE of pC is less dependent on the mean value than is the case for EC.

EDGE EFFECT  If a plot is bordered by a FALLOW path, or a different crop, the growth near the edge may be atypical of the treatment (sometimes it's better, e.g. cereals without N, sometimes worse, e.g. beans attacked by aphids). Cf. COMPETITION EFFECT.

EFFECT  A straightforward word but often we assume that effect = main effect. This is the difference in mean yield produced by the FACTOR considered, averaged over all LEVELS of other factors in the experiment and in the appropriate units – usually 'lb (or kg) per plot' or more often the conventional unit of yield per unit area (e.g. cwt per acre). If the factor concerned has three or more levels more than one figure is needed to describe the effect; there is no unique way of calculating these figures. E.g. for levels 0, 1, 2 of a factor $A$ we may use

|  | $a_1 - a_0$ | (response to first unit) |
|---|---|---|
| and | $a_2 - a_1$ | (additional response to second unit) |
| or | $a_2 - a_0$ | ('linear response') |
| and | $a_0 - 2a_1 + a_2$ | ('quadratic response') |

(Note the convention – the factor denoted by a capital, the 'levels' by lower case.)

Unfortunately some people (perhaps those who spend more time in the theory than in the practice of statistics) use the word 'effect' in a different sense. The phrases 'fixed effects' and 'random effects' give a warning that such people are abroad. 'Effect' in this sense is equivalent to 'treatment deviation' (or 'block deviation' etc.) in my Chapter 11. The word is also used (a little loosely) to indicate the estimates of these deviations. This usage is confusing, for example if you have a factor $A$ at two levels (0 and 1) and it gives mean yields

$$\bar{y}_0 = 10$$
$$\bar{y}_1 = 20$$

the 'main effect' of $A$ = +10 and many people (I included) will say 'the effect of $A$ is +10' but the other school says 'effects of $A$: level 0, −5; level 1, +5'.

ERROR  Not to be confused with 'mistake' let alone 'sin'. Used loosely as equivalent of RESIDUAL (sense 1). But the phrase 'gross error' bridges the gap between statistical and common usage (see Appendix B).

EXPERIMENT  Anything that happens in a field which allows comparison of two or more differently-treated areas constitutes a field experiment. The word 'trial' is much used (especially in the comparison of varieties) but carries a faint suggestion that someone is 'trying out' something recently invented. I think we are going to live with both words in common use, more or less as alternatives. An 'observation study' is an experiment with inadequate replication.

FACTOR, FACTORIAL  An experiment is factorial if it includes all combinations of at least two factors (but see note on fractional replication under REPLICATION). Each factor may be represented by two or more levels. A very simple factorial experiment might have two varieties and two levels of nitrogen fertiliser $- 2 \times 2 = 4$ treatments. Such an experiment could be fitted into several DESIGNS, e.g. randomised BLOCKS or LATIN SQUARES. Strictly one uses capitals for the factors ($A, B,$ etc.) and lower case for the levels ($a_1, a_2, b,$ etc.). Don't be alarmed by the use of (1) for the 'untreated' as in (1), $a, b, ab$; this shows a decent respect for group theory. Factorial experiments are known as '2 x 2', '3 x 2 x 2', etc., with contractions such as $3^3$.

FALL  (U.S.A. and Canada) = autumn (U.K.) but 'fall wheat' = 'winter wheat'.

FALLOW  Land that carries no useful crop is said to lie fallow. Fallows can be 'bare', that is cultivated or sprayed to kill weeds, etc., as they appear, or not, that is weedy. Weedy fallows were the third COURSE in Saxon agriculture, the three fields surrounding a village carrying autumn-sown CORN, spring corn and fallow in rotation. These fallows were probably weedy enough to give some grazing for livestock and they gained fertility from the droppings. In Iraq in a similar three-field system I have seen a fallow that had been irrigated to increase the growth of weeds for the stock.

FARMYARD MANURE  (U.S.A. Barnyard manure, BYM) In damp temperate climates farmyard manure (FYM, muck or dung) is usually made with plenty of straw (to sop up the moisture from the dung and urine and save the beasts, usually cows, calves, bullocks or pigs, rarely sheep, from getting too wet and dirty). If the straw has not rotted much the manure is 'long' or 'fresh'; later it is 'rotted' or 'short'. Sometimes it is carted from where it is made and 'clamped' until required; this ensures good rotting.

In arid hot climates little or no straw is added deliberately but wasted feeding stuffs sometimes act as a substitute. The manure has much less moisture than in cool damp places and is often dusty to handle. Such stuff can be more reasonably called 'dung' than the strawy stuff referred to earlier.

FERTILITY TREND (or FERTILITY DRIFT) The tendency for plots of the same treatment to yield more at one side of an experiment, less at the other. Coped with by division into BLOCKS (perhaps with CONFOUNDING to lessen block-size), or by taking out a (usually linear) regression on position. The latter is mildly deplorable as it leaves us with two alternative interpretations of the same data. Prevention being better than cure, if a fertility trend is suspected when an experiment is planned, use a suitable design (e.g. LATIN SQUARE or randomised blocks with confounding to lessen block-size).

FYM  See FARMYARD MANURE.

GRAND MEAN  See MEAN.

GUARD-ROW  A row of crop at the side of a plot which, though it receives the treatments applied to the plot, is not included in the area harvested for the estimation of yield. The growth of guard-rows may be atypical because of EDGE EFFECT (e.g. if bare ground adjoins the plot) or COMPETITION EFFECT. In an experiment where different crops are grown on neighbouring plots, the guard-rows are a useful buffer against spray-drift. Occasionally more than one guard-row may be needed at the edge of a plot.

HEADLAND  Originally, a strip of land next to the hedge on which the ploughman turned his horses. Now we have field headlands for the turning of tractors with all sorts of implements (e.g. in a potato field a headland may be left bare for turning sprayers, etc.). Experiments often have headlands (perhaps well inside the field) — some are 'internal', e.g. for turning the drill between rows of plots if (say) different varieties are randomised in the experiment. The drill dodges about on the headlands and drills all plots of variety A before cleaning out the hopper for variety B. Occasionally an experiment will need 'sidelands' as well as headlands — for cross cultivation or for spraying across the rows. A headland may be bare FALLOW or sown with a convenient crop, the decision depending on the uses of the headland.

INCOMPLETE BLOCKS  Describes designs in which each BLOCK contains less than all the treatments included in the experiment. For a special case, see LATTICE. (FACTORIAL experiments with CONFOUNDING are an example of incomplete blocks.) Balanced incomplete block designs have the property that each pair of treatments occurs together in the same number of blocks as any other pair of treatments.

INFORMATION  'Information' is a scale of measurement with
arbitrary units. The information supplied by an experiment on the
comparison of any two treatments is inversely proportional to the
VARIANCE of the mean difference between the two treatments.
Other things being equal, doubling the replication doubles the
information. Contrast PRECISION.

INTERACTION  A numerical measure of the degree to which the
variation of one FACTOR alters the effect of another. For two
factors at two levels, $A$ and $B$ (say), the interaction $A \times B$ ($A . B$ or
$AB$ for short) $= \frac{1}{2}$ mean $(a_2 b_2) + \frac{1}{2}$ mean $(a_1 b_1) - \frac{1}{2}$ mean $(a_1 b_2) -$
$\frac{1}{2}$ mean $(a_2 b_1)$. For factors at more levels (or for more factors) things
are more complicated, but for all interactions in a $2 \times 2 \times 2 \times \ldots$
scheme of treatments the divisor ($\frac{1}{2}$ in the example) is such as to
make the interaction in effect the difference between 2 means,

$$\frac{1}{n} (\text{total of } n \text{ plots}) - \frac{1}{n} (\text{total of } n \text{ others})$$

This implies (in the normal $2^n$) that main EFFECTS and all
unconfounded interactions have the same STANDARD ERROR.

There is a convenient algebraical shorthand for effects and interactions; the interaction $AB$ given above may be written

$$\frac{1}{2} (a_1 - a_2)(b_1 - b_2)$$

If you expand this and interpret the 'products' $a_1 b_1$, $a_1 b_2$, etc.,
as the mean yields of the corresponding treatments this is equivalent
to the definition above.

Similarly

$$\frac{1}{2} (a_1 + a_2)(b_1 - b_2)$$

represents the main effect of factor $B$ (averaging the two levels of
factor $A$). This extends naturally to cases like

$$\text{linear } A \times \text{quadratic } B = \frac{1}{Q} (a_2 - a_0)(b_0 - 2b_1 + b_2)$$

but the divisor $Q$ in such cases can be controversial — my rule is to
choose $Q$ so that the units of the interaction are those of a difference
between two means (my $Q = 4$ in this example).

*Warning.* What I have written above is based on what I read about
25 years ago in Yates's Technical Communication[8]. I have discovered
recently that there are two other conventions about divisors
appropriate for calculating interactions. I am not going to confuse
you by giving these — or even telling you how to look them up. My
convention is of course the best.

The only instance where it does not work is the rare one of an interaction that is partly confounded but can be estimated by adding to the usual formula a block-correction; here you should calculate the divisor ignoring the block-correction. (A different divisor is required in calculating the corresponding standard error.) An example may help:

Factors $A$, $B$ give four treatments $(1)$ $a$ $b$ $ab$. Experiment in two blocks of two, confounding $AB$:

Block I    $a$    $b$
II    $(1)$    $ab$

Now add a plot to each block, each with an additional treatment $x$; call their yields $x_1$ $x_2$

Block I    $a$    $b$    $x_1$
II    $(1)$    $ab$    $x_2$

Then the interaction $AB$ may be estimated from a modified formula $(1) -a -b + ab -2x_2 + 2x_1$. (This is orthogonal to blocks and to factor $A$, factor $B$ and to the comparison between $x$ and the factorial treatments.) The divisor for the estimate of $AB$ is 2 (as it would be in an unconfounded experiment without treatment $x$) but the divisor needed for calculating the standard error of the estimate is

$$\sqrt{[1^2 + (-1)^2 + (-1)^2 + 1^2 + (-2)^2 + (2)^2]} = \sqrt{(12)}$$

INTERFERENCE  Any effect of a treatment, which is not eliminated by the use of GUARD-ROWS, on plots in the same layout receiving other treatments. Most likely to occur in experiments on foliage diseases and pests.

LATIN RECTANGLE  A 'doubly-restricted' design in rows and columns — the number of rows not being the same as the number of columns. Healy[32] shows the usefulness of latin rectangles of size 4 x 8 (32 plots in all) for $2^3$, $2^4$ and $2^5$ experiments. In the $2^3$ each row of eight plots is a replicate but the three-factor interaction is confounded with columns. In the $2^4$ and $2^5$ certain interactions are confounded with rows, others with columns. These designs, like the LATIN SQUARE, eliminate a linear trend of fertility in any direction.

LATIN SQUARE  Describe method of RANDOMISATION with double restriction — each TREATMENT occurs just once in every 'row' and once in every 'column'.

A B C D
B C D A
C D A B
D A B C  is a 4 x 4 latin square.

In field experiments 'rows' and 'columns' usually have their obvious meanings, but not necessarily. If an experiment has to be laid down with plots in one row on a site where a marked FERTILITY TREND is expected, a Latin square (e.g. *ABCCABBCA*) may be the best design — you will find that the centres of gravity of the plots of each treatment all coincide. This design therefore is unaffected by a linear fertility trend (and by the linear component of a more complex trend). If a Latin square is used in an unconventional LAYOUT such as this it must be analysed as a latin square. The DESIGN not the layout determines the method of analysis.

The 3 x 3 square just used is a disreputable design (having only two DEGREES OF FREEDOM for ERROR). The 4 x 4 square (6 d.f. for error) is sometimes an attractive alternative to four BLOCKS of four (with 9 d.f.) if a trend of uncertain direction is feared.

LATTICE  Lattices are a special case of INCOMPLETE BLOCK designs. In a lattice the number of treatments is an exact square (but remember that you can duplicate one or more treatments to make up the number to an exact square).

In a balanced lattice each pair of treatments occurs together in a block the same number of times in all.

LATTICE SQUARES  A class of designs useful for comparing large numbers of varieties; like latin squares they eliminate variations of fertility, etc., in two directions. A set of balanced lattice squares has the following properties:

(1)  Each treatment occurs exactly once in each square.
(2)  Each pair of treatments occurs together in the same row or column of a square the same number of times in all.

Such sets exist for 9, 16, 25 (not 36) 49, 64 . . . treatments.

LAYOUT (or ARRANGEMENT)  The geometrical relation of the plots of an experiment on the field. Distinct from the DESIGN and from the RANDOMISATION. Includes size of plot, orientation, HEAD-LANDS, etc.

LEACH  When water percolates down through the topsoil (and perhaps the subsoil too) the soil is said to be leached. If on its journey it carries nutrients in solution, these are also said to be leached.

LEVEL  A FACTOR is said to be applied at 2, 3, 4, . . . levels. Where the factor is (say) a fertiliser, levels are rates per unit area. Two levels may be 'absence' and 'presence' or single and double rates, or something more complicated. If a factor such as variety of crop is included it seems rather unnatural to use the word 'level' but there is no easy alternative. Levels are often listed with $v.$ (= versus) between them, e.g. 'the comparison $(n_0$ v. $n_1)$' or '$(n_0$ v. $n_1$ v. $n_2)$'.

LEY  Grass to be left for one or more seasons but eventually to be ploughed up. This brief description needs much amplification. 'Grass' means any crop or combination of crops that can be grazed by livestock or cut for immediate or later feeding to livestock. Leys are often composed of several species of grass perhaps with several strains of red and white clover; on the other hand a ley may have one species only (e.g. LUCERNE). Leys may be sown 'in the open' or 'under a nurse-crop' (i.e. undersown). The nurse-crop may be a cereal (commonly barley) that is harvested ripe or a fodder-crop (e.g. rape) that is grazed off by livestock. Undersown leys come into full production in the season after sowing; leys sown in the open, if of suitable strains, can produce much bulk in the year of establishment. 'Ploughed up' should be taken as including 'destroyed by chemical means, e.g. by paraquat'.

LUCERNE  (U.K.) = Alfalfa (U.S.A.) = *Medicago sativa.*

MEAN  Usually equivalent to 'average' (i.e. arithmetic mean). 'Mean' is perhaps a little more general than 'average' ('geometric average' is hardly acceptable). 'Treatment-mean', 'block-mean', need no comment. 'Grand mean' is used of the average of all plots of an experiment. 'Weighted mean' indicates a mean such as

$$\bar{y}_w = (y_1 w_1 + y_2 w_2 + \ldots + y_n w_n)/(w_1 + w_2 + \ldots + w_n)$$

$w_1, \ldots, w_n$ being the 'weights' (in the statistical sense). If all of $w_1, \ldots, w_n$ are equal $\bar{y}_w$ simplifies to $\bar{y}$, the (unweighted) mean.

MICROPLOT  A mongrel word but serviceable. In general any very small plot often of 20 sq yd or less, but size is not the best criterion. 'Microplot' is best used of a plot so small that normal field methods of cultivation, sowing, harvesting cannot be applied; microplots may be dug by hand or cultivated by rotary hoes, sown by hand or by a single-row drill, harvested by hand.

MUCK  A synonym of FARMYARD MANURE.

ODDS AND ENDS (O & E)  The bits of a field left over after an experiment (or experiments) has been laid down. Distinct from GUARD-ROWS and DISCARD.

PATH  Down-paths (also called 'side-paths') are parallel to the rows of a crop, cross-paths (more or less) at right angles. The former are usually left blank at drilling, the latter cut later by rotary hoe or motor scythe. Path widths (usually 1–3 ft) are shown on plans. Plot dimensions usually exclude paths; measurements over several plots include internal paths but not extreme ones. For all crops in rows, my convention is that the stated plot width includes a half-row space outside each edge row so that (e.g.) in cereals sown at 7 in. the space between extreme rows of neighbouring plots = the path width + 7 in.

pC  See EC.

pH  pH = $\log_{10}$ (volume in litres which contains 1 g of $H^+$ ions). This applies to a solution. Determination of the pH of a soil depends on the (more or less arbitrary) rules by which you obtain a solution. Most crops prefer pH's above 5.5 or 6.0. Some (especially legumes) seem to do best at pH about 6.5 to 7. The pH of soil in the field changes with time and up-to-date determinations are often needed before experiments are laid down.

PHASE  Plots which carry the same sequence of crops but which are out of step by one, two, etc., years are in different phases. The term 'phase' is also used to denote a particular component of the cropping CYCLE. All the best experiments have all phases equally represented.

PLAN  The final drawing of a field experiment, usually drawn *after* the experiment has been laid down. (Contrast SKETCH.) It is often better not to draw to scale, but to use different scales in the two directions and indicate PATHS and perhaps HEADLANDS by single lines. If you can, preserve the true alignments of the PLOTS. The dimensions of each plot should be given and the overall dimensions; I normally *exclude* paths and headlands at the edges of the experiment. A north point should be given and perhaps another reminder of the orientation, e.g. an arrow indicating the direction of the 'Red Lion'. Other information you may like on your plans depends on circumstances. The following suit mine:

Treatments, Design, Variety, Basal manuring, Area harvested, Field name, Special requirements (e.g. 'no weedkiller to be applied').

PLAT  (U.S.A. to about 1930) = PLOT.

PLOT  A piece of land, to be compared with another. The hierarchy Block/Sub-block/Plot/Sub-plot is arbitrary to some degree, e.g. an experiment in blocks with factor $A$ on whole plots, factor $B$ on sub-plots can equally be described as in blocks, with $A$ on sub-blocks, $B$ on plots.

PLOT FACTOR  The number by which you multiply the recorded yield of a single plot to turn it into the desired final units (e.g. kg per hectare). The two-plot factor is half the one-plot factor, and so on.

PLOUGH (= U.S.A. plow)  In the U.K. 'plough' unqualified means 'mouldboard plough'. Disc-ploughs are little or not at all used in the U.K. As far as I know all non-reversible mouldboard ploughs and disc ploughs move soil to your right as you follow the plough. It is probably best to reserve the word 'plough' for an implement which to a greater or less extent turns the soil over. 'Chisel-ploughs' are better called 'deep-tine cultivators'.

Reversible ploughs (correctly so-called, often called 'one-way ploughs') can turn furrow(s) to right or left by choice of one or other of two sets of bodies. Thus a field can be ploughed 'one way', i.e. all furrows are turned in one direction without open furrows or 'gathering ridges'. These were more or less unavoidable with earlier methods of ploughing in 'lands', and often made the choice of sites for experiments difficult. Primitive wooden 'ploughs' without mouldboards are still used in some countries and have some of the advantages of the reversible plough.

PLOW (U.S.A.) = plough (U.K.)

PRECISION  Two experiments (differing perhaps in DESIGN, LAYOUT, etc.) testing similar treatments applied to the same crop are said to be of differing precision if the STANDARD ERROR per plot is greater in one than in the other (the latter has greater precision).

PREPARATORY CROP  In some CROP-SEQUENCE experiments preparatory crop(s) may be grown to produce suitable conditions for TREATMENT CROPS (e.g. to produce sites with much or little soil-borne disease). Yields are not normally taken from preparatory crops.

RANDOMISE, RANDOMISATION
    (1) The process of compiling random orders of treatments with the restrictions proper to the DESIGN used.
    (2) One particular result of this process is a randomisation.
    The logical (and chronological) order is:

    (1) design.
    (2) Randomisation.
    (3) LAYOUT.

For 'restricted randomisation' see Grundy and Healy[31], and Dyke[23].
    Producing many randomisations takes much time, especially if many PLOTS or CONFOUNDING are involved. At least one

electronic computer program is available to produce randomisations (including restricted ones) for many of the designs commonly used.

**RANDOMISED BLOCK** See BLOCK.

**RECOVERY** See UPTAKE.

**REPLICATE, REPLICATION** In a non-FACTORIAL experiment, a replicate means the set of all the TREATMENTS that are included. In a FACTORIAL experiment a replicate means the set of all combinations of all factors at all LEVELS. All these combinations need not appear in the experiment, e.g. a valid experiment can be done with six factors each at two levels (a '$2^6$' experiment) using only 32 (= $2^5$) plots. This is a 'half-replicate', an example of 'fractional replication'.

'There is no replication' means (paradoxically) that there is just one replicate.

**RESIDUAL** An overworked word. I find three usages relevant to field experiments, one of which I deplore.

(1) If you regard the yield of a plot as the sum of a number of bits (some may be negative) — mean of all plots, block-deviation (i.e. deviation of block-mean from the grand mean), treatment-deviation, deviation due to regression on plant number . . . the bit you have not explained is the residual (or, more exactly, 'plot-residual'). Many electronic computer programs print residuals for inspection; in this they are superior to human computers who seldom have the patience to calculate them. If you make a habit of scanning arrays of plot-residuals you may like to know the rule for calculating their STANDARD ERRORS. In a simple experiment laid down in '$r$' randomised BLOCKS each containing a replicate of '$t$' treatments with $rt = n$ plots in all, with neither CONFOUNDING nor plot-splitting, the standard error of each plot-residual is

$$\sqrt{[(\text{Sum of squares for error})/n]}$$

This gives you a means of picking out the plots whose residuals are suspiciously large (e.g. you may mark all that are greater than twice (or three times) their standard error. This may help you to detect a gross error, or a FERTILITY TREND; but be cautious for two reasons (*a*) if you are looking at 20 residuals you may expect about one to reach 'significance' at the 5% level by pure chance (with 100

residuals, one 'significant' at the 1% level and so on), (b) the
residuals, from their method of calculation, are correlated
one with another. The formula given above also applies to a
LATIN SQUARE but if the experiment involves confounding or plot-splitting things are more complicated.
(2) A FACTOR applied once for all has a DIRECT effect on the
first crop, a residual effect on subsequent crops. Occasionally
'residue' is used, e.g. 'the residue of superphosphate'.
(3) 'Residual herbicide' means a herbicide (I prefer the word
'weedkiller' as simpler and more precise) that continues to
kill or suppress weeds some weeks after application (but, it
is hoped, does not adversely affect the present or any subsequent crop). 'Persistent' or 'soil-acting' seem appropriate
adjectives that avoid the awkward ambiguity of 'residual'.

RESPOND, RESPONSE  If a TREATMENT changes the YIELD of a
crop the crop is said to 'respond' to the treatment. 'Response to' =
'EFFECT of' but the two phrases are (by usage) not always inter-
changeable. 'Effect of a new variety' (not 'response to') and see
RESPONSE CURVE (not effect curve!).

RESPONSE CURVE  The graphical representation of the effect on
crop-yield of increasing the rate of application (usually of a
fertiliser). Although long discredited (at any rate for N on cereals)
the Mitscherlich asymptotic curve (Crowther and Yates[21]) is still
probably the most-used mathematical model. If the rates of application are evenly spaced, the further response to each unit is in a fixed
proportion (between 0 and 1) to the predecessor. The curve steadily
approaches a limiting value (but never reaches it). In real life it seems
that some response curves begin to decline at some point but no
generally accepted formula has this property.

ROTATION  A sequence of crops that may be repeated indefinitely.
Rotation Experiment (strictly, Cyclic Rotation Experiment): any
experiment designed to run through several CYCLES. For the whole
truth on rotation experiments see Patterson[42].

ROUNDING OFF  See Appendix A for a discussion of rounding off
yields and figures calculated from yields.
    In calculating amounts of fertilisers, etc., to be applied to plots
it is usually enough to round to the nearest 1%–3%. Personally I
work out and round off the smallest amount and then multiply this
by 2, 3, etc., if several rates are being compared.

SCORE  A term applied to any record of the crop or soil of a plot not
made by direct measurement but by comparison with a more or
less arbitrary scale. Plots may be scored for crop density, colour,

height or incidence of lodging or disease. Sometimes a key may be carried by the scorer (e.g. for common scab on potato tubers) but often the scale is established in his mind by a preliminary inspection of some of the plots. Scoring (ideally) should be done without knowledge of the treatments of the plots, and the plots of a BLOCK should be scored consecutively.

SERIES  In ROTATION experiments, plots in the same PHASE of the cropping rotation(s) form a series. Each series may contain one or more REPLICATES of the TREATMENTS which may be arranged in BLOCKS or for that matter, LATIN SQUARES.

SIDELAND  See HEADLAND.

SKETCH  Drawing of an experiment made before it is laid down. The relation of the sketch to the final PLAN may be close — or less close.

STANDARD  See BASAL.

STANDARD ERROR (S.E.). (& S.E. % = Coefficient of Variation). The simplest measure of the 'spread' of variation due to random (i.e. unassigned) causes. To be preferred to 'Least Significant Difference' which (unlike the standard error) depends on (1) the assumption of a 'normal' (Gaussian) distribution, (2) the chosen level of likelihood deemed 'significant'. In a field experiment distinguish between 'S.E. per plot', 'S.E.%' (= 100 × S.E. per plot/mean yield) 'S.E. of treatment mean' and 'S.E. of treatment-effect' or 'S.E. of difference'. A good analysis gives S.E. per plot (actual and %) as well as S.E.s of treatment means. 'Standard deviation' (S.D.) is to a population as 'Standard error' is to a sample. S.D. (which in experiments, etc., we never know) is the square root of VARIANCE.

SURROUND (noun)  The area of land close to an experiment. An experiment may need a sown surround (e.g. to lessen the risk of damage by birds at the edge) or a FALLOW surround (e.g. to allow a good chance for drifting aphids to alight on the crop).

TEST-CROP  See TREATMENT-CROP.

TILL, TILLAGE  To till (the soil) = to cultivate. To till (a crop) = to prepare the land and sow. For 'tillage land' see ARABLE.

TOP DRESSING  Used of a fertiliser applied to a growing crop; the common way of applying N to winter wheat in the U.K., where mild wet winters usually (but not always) cause N applied in autumn to be LEACHED before the crop can assimilate much of it.

TREATMENT  A treatment is something to be compared with another treatment. In a FACTORIAL experiment every different combination of the factors is a treatment, e.g. a 2 x 2 experiment has four treatments. Contrast BASAL.

TREATMENT-CROP and TEST-CROP (In experiments where the treatments are different crop rotations or sequences.) Treatment-crops are those grown when the plots carry different crops; in any later season, if all plots carry the same crop, this is referred to as a test-crop.

TRIAL  See EXPERIMENT.

UNIFORMITY TRIAL (It ought to have been called 'variability trial' but it's too late now.) A piece of field work that involves everything done in an EXPERIMENT, except the application of TREATMENTS. Usually a large number of PLOTS are marked out (before or after sowing the crop) and all are treated identically throughout. The YIELD of each plot is recorded separately. Such data can be analysed in many ways; you can put on more or less any DESIGN you care to imagine (e.g. RANDOMISED BLOCKS, LATIN SQUARES) and analyse the yields accordingly. For instance, if there are 64 plots in eight rows of eight you can analyse as for:

    (1)  blocks of 8 taken row-wise;
    (2)  blocks of 8 taken column-wise;
    (3)  blocks of 4 (2 to a row);
    (4)  an 8 x 8 Latin square;

or for that matter (discarding some plots)

    (5)  blocks of 7;
    (6)  a 7 x 7 latin square,

etc., etc.

A typical analysis [for example (1) above] is

| Blocks | 7 d.f. | |
|---|---|---|
| Error | 56 | |
| Total | 63 | (1) |

There being no treatments, you can't take out a sum of squares for them. You could of course put in dummy treatments duly randomised to get

| Blocks | 7 | |
|---|---|---|
| 'Treatments' | 7 | |
| Error | 49 | |
| Total | 63 | (2) |

but there is no point in doing this.

The sum of squares for 'Treatments' plus Error in (2) is equal to the sum of squares for Error (56 d.f.) in (1) and, averaging over all possible randomisations, the mean mean square for 'Treatments', and the mean mean square for Error (49 d.f.) are both equal to the mean square for Error in (1). Error (1) is a sample of 56, and Error (2) a sub-sample of 49, from the same imaginary infinite population.

If you analyse the same set of data in several different ways (e.g. 64 plots as above can be analysed as 8 blocks of 8 and again as an 8 x 8 latin square) you can calculate the error mean squares and from them the PRECISION of any one design relative to any other.

Note that the error mean square in the analysis of a uniformity trial has the nature of within-block error, *not* treatments x blocks. So, strictly, uniformity trials give an idea of the magnitude of error variances of experiments *only* if there are no appreciable treatment x blocks interactions, i.e. if treatment differences and block differences are additive (see Section 11.7).

UPTAKE  If a fertiliser containing say N is applied to a crop the harvested produce will (usually) contain more N than if no N had been applied. The difference is commonly called the 'uptake' of N; sometimes it is necessary to distinguish between 'apparent uptake' (the difference just defined) and 'true uptake'. The latter is the amount of the fertiliser-N found in the harvest produce. This can be measured by the use of 'labelled' fertiliser, e.g. ammonium sulphate made with the artificial isotope of nitrogen $^{15}$N. If the fertiliser stimulates the growth or activity of the roots the treated crop may 'explore' the soil more efficiently and the apparent uptake will be greater than the true uptake. It is occasionally possible to find an apparent uptake greater than the amount of the nutrient applied in the fertiliser.

The word 'recovery' is also used in the sense of uptake; 'the crop recovered 10 per cent of the applied N' is equivalent to 'the (apparent) uptake of N was 10 per cent'. 'Uptake' is usually given as weight per unit area, 'recovery' as per cent of amount applied.

VARIANCE  The mean square deviation from the mean (usually denoted by $\sigma^2$). The variance is a function of the population; it is estimated from samples by means of the formula

$$s^2 = [1/(n-1)] \sum_i (y_i - \bar{y})^2$$

which is shorthand for

$$[1/(n-1)] [(y_1 - \bar{y})^2 + (y_2 - \bar{y})^2 + \ldots + (y_n - \bar{y})^2]$$

$[y_1 \ldots y_n$ are the values in the sample, $\bar{y} = \frac{1}{n}(y_1 + y_2 + \ldots + y_n)$, is the sample mean]. In analysis of variance the mean square in each line is calculated from slightly more complicated formulae of this type. On the 'null hypothesis' (that the yields from different blocks, treatments, etc., all come from the same population) all are estimates of $\sigma^2$, the population variance.

Variance is a pretty fundamental idea and I think it's worth trying to illustrate it at length.

Suppose I buy two packets of lump sugar and take four lumps from each, weigh them and record their weights in grams as follows:

| Packet A | 10 | 9 | 10 | 11 |
|----------|----|---|----|----|
| Packet B | 8  | 7 | 11 | 14 |

For each packet the total weight of the four sample lumps is 40, and the mean is 10. But it is clear that the sample from packet B is more variable than sample A; the range of variation is 2 for sample A, 6 for sample B. On the evidence that we have, the lumps in packet B appear to be more variable in weight than those in packet A. A quantitative measure of variability would be useful — that is, we should like two numbers to compare, one representing the 'variability' of each sample. The 'ranges' just given are obviously such measures but on examination they turn out to be not very useful. For example if we enlarge each sample by taking extra lumps one at a time from each packet, the range of each sample will increase occasionally. Also the range, being calculated from only two members of the sample (the lightest and the heaviest), varies widely from sample to sample (e.g. if we take a second sample from packet B we might by luck get a much smaller range).

A more useful measure is the 'sum of squares of deviations' or simply 'sum of squares' or 'SS' for short. This is calculated as follows (taking sample A):

| Sample values | 10 | 9 | 10 | 11 | (mean 10) |
|---------------|----|---|----|----|-----------|
| Mean | 10 | 10 | 10 | 10 | |
| by subtraction, deviations from mean | 0 | −1 | 0 | +1 | (total 0) |
| Sum of squares = | $0^2 + (-1)^2$ | | $+0^2$ | $+1^2 = 2$ | |

For sample B we have

| Sample values | 8 | 7 | 11 | 14 | (mean 10) |
|---------------|---|---|----|----|-----------|
| Deviations | −2 | −3 | +1 | +4 | (total 0) |
| Sum of squares = | $+(-2)^2 + (-3)^2$ | | $+1^2$ | $+4^2 = 30$ | |

So that (if we use 'sum of squares' as a measure of variability) sample B is 15 times more variable than sample A.

Note that the sum of squares depends on all the observed values, and would in general be changed if any one value were changed, whereas the range depends on the two extreme values only. Note also that similar calculations could be made if the two samples had different means. Indeed if the readings of sample B were

|  | 18 | 17 | 21 | 24 |
|---|---|---|---|---|
| or even | 108 | 107 | 111 | 114 |
| instead of | 8 | 7 | 11 | 14 |

the sum of squares for sample B would still be 30 — you should check this by calculation.

But what if we wish to compare two samples that contain different numbers of sample values? For example we might have:

Sample A  (4 values)   10    9   10   11
Sample C  (6 values)    9   12   11    8    7   13.

Now the totals are different (40 and 60 respectively) but the means are the same, 10. But obviously the sums of squares are (like the totals) not directly comparable; in fact every additional value we may add to an existing sample, unless it exactly equals the mean of the sample so far, inevitably increases the sum of squares. We might guess that we ought to treat the sum of squares as we treat the total to obtain the mean, that is, divide it by the number of values in the sample. In the event this does not answer as well as dividing by one less than the number of values (3 for sample A, 5 for sample C in our second example). The result is called the 'mean square deviation', 'mean square' or 'sample variance'. This odd procedure should perhaps be justified. Here are one or two lines of thought worth trying.

First, a sample of one value gives you *some* information about the mean of the population (in fact, if you can't get any more information you must logically use the sample value as your estimate of the population mean). But it gives you *no* information at all about the variability of the population. A sample of two gives a little information about the variability of the population (you have one 'comparison' available for the estimation of variance). A sample of three gives twice as much information as a sample of 2 because you now have two independent comparisons available. (For the idea of 'comparisons' see Chapter 11.) The divisor 2 which is appropriate to the sum of squares of deviations of a sample of three corresponds to the averaging of two independent estimates of variance, each derived from one degree of freedom.

Next, consider what happens when we subsample from a sample

already taken. Take for example the four values we used above as our original sample 'B', viz.,

$$8, \quad 7, \quad 11, \quad 14$$

If we take subsamples of two there are exactly six possible subsamples. These are:

$$(8, 7) \quad (8, 11) \quad (8, 14) \quad (7, 11) \quad (7, 14) \quad (11, 14)$$

If the subsampling is done by choosing two at random from the four values each of these six pairs is equally likely to turn up. Now calculate for each subsample the mean and sum of squares of deviations from this (subsample) mean. For the first subsample the mean is 7.5, the deviations are $+0.5$, $-0.5$ and the sum of squares is $(+0.5)^2 + (-0.5)^2 = 0.5$. In this case, the divisor being 1, the mean square is 0.5. If you do similar calculations for the other five possible subsamples (which I exhort you to do, as an exercise) and average the six mean squares so obtained the result is (barring mistakes) 10 — which as we saw above is the mean square deviation derived from the sample of 4. Note that if you use as divisor the number of values in the sample or subsample (instead of one less) this neat equality does *not* hold.

[By the way, if you have done the exercise, you will have got tired of squaring and adding pairs of numbers which are equal but opposite in sign, e.g. $(+0.5)^2 + (-0.5)^2$, and will be glad to use the short cut based on the formula

$$\left(x_1 - \frac{x_1 + x_2}{2}\right)^2 + \left(x_2 - \frac{x_1 + x_2}{2}\right)^2 = \frac{(x_1 - x_2)^2}{2}$$

You will find more on the same lines in Chapter 11.]

Now we consider the variance of the mean of the observations in a sample. The general formula is:

variance of mean of $n$ values $= \dfrac{1}{n}$ (variance of population).

I am not going to offer a proof of this and I do not find it easy to give a convincing example. Here is the best I know.

Consider a population of 1's and 2's (in equal proportions — you would get such a population by repeatedly tossing an unbiased coin and recording each head as 1, each tail as 2). The variance of the population is the mean square deviation from the mean (which is 1.5). Every deviation is $+0.5$ or $-0.5$ so the variance is $(0.5)^2 = 0.25$.

Now consider samples of two drawn from this population; we shall calculate the variance of the means of such samples. On average one sample in every four will be $(1, 1)$, two will be $(1, 2)$

(we neglect the order of drawings) and one will be (2, 2). The corresponding sample means are 1.0 (once in four times), 1.5 (twice in four) and 2.0 (once). The mean of the population of sample means is of course 1.5 and the deviations are −0.5 (once), 0 (twice) and 0.5 (once). The mean square deviation of the sample means is therefore

$$(-0.5)^2 \times (\tfrac{1}{4}) + (0)^2 \times (\tfrac{2}{4}) + (0.5)^2 \times (\tfrac{1}{4})$$

which equals 0.125.

This is the variance of the population of sample means and it is equal to half the variance of the original population, in pleasant agreement with our formula, $n$ being 2 in this case.

Finally, a paragraph on sampling from a finite population, e.g. all the lumps of sugar in a certain packet. Suppose there are exactly 100 lumps. If we weigh every lump and calculate the mean, this is an exact value *not* subject to variability − in fact its 'variance' is zero. Similarly, the mean of 99 values (perhaps we put one in our tea before finishing the weighing!) has less variance than the mean of a sample of 99 from an infinite population. This leads to the idea of a correction factor for samples from a finite population:

(variance calculated as if population were infinite) $\times (1 - f)$ where $f$ = the proportion of the population that are in the sample.

This formula is worth remembering (and, rarely, worth using) when you take several replicate samples from each plot of an experiment which in total account for an appreciable proportion of the plot. It may be applied to the *sampling* error of the mean estimate for each plot, but *not* to the 'between-plots' error; this needs *no* 'correction'.

WEIGHTED MEAN  See MEAN.

YEAR  Trickier than it sounds. In England wheat sown in October 1970 will be harvested about August 1971. I call this 'wheat 1970/71' or (more often) 'wheat 1971'; any job done on that crop is by convention referred to as done in 1971. The Ministry of Agriculture Fisheries and Food publish statistical summaries in which '1970/71' refers to crops harvested between 1 June 1970 and 31 May 1971. So my 'wheat 1971' is to them 'wheat 1971/72' *not* 1970/71.

YIELD  Unless qualified usually means 'yield per unit area of the main product of the crop', e.g. for cereals, grain at 85% dry matter (yield of straw would be stated explicitly). In our present context, when most of us feel that the world's population is increasing out of proportion to its agriculturally productive area, yield per unit area seems the critical measure − almost anything that increases it must be good, anything that decreases it bad. But long ago when men were few and acres many yield per unit quantity of seed sown was the measure naturally used. Anyone who has sat hungry through a winter trying not to eat the seed-corn set aside for next season is likely to think in terms of yield per pound (or ounce!) of seed. (See the parable of the sower, Mark[38].)

Some farmers now, not unnaturally, think in terms of yield per unit cost – hence the dreary references to 'gross margin' that clutter some journals. In one of my waking nightmares I see a future when men have squandered most of the world's resources of oil and coal and the critical measure is yield per unit of mechanical work done. A minor compensation of that post-golden age is that if harvested produce is expressed in terms of energy (as in the use of calories by dieticians) 'yields' will be dimensionless and independent of the units used (joules per joule, calories per calorie and so on).

# BIBLIOGRAPHY AND REFERENCES

Books on methods of field experimentation are few; most of them have been published by departments or ministries of agriculture, experimental stations or companies involved with fertilisers or pesticides. The ones I have found useful are listed below, with comments.

## 1. BOOKS ON FIELD METHODS

1. Lynch, P. B., *Conduct of Field Experiments*, Bulletin No. 399, New Zealand Department of Agriculture, 154 (1966)
   The most up-to-date and comprehensive book I have found. Especially good on grassland experimentation.

2. Pearce, S. C., *Field Experimentation with Fruit Trees and Other Perennial Plants*, Technical Communication No. 23, Commonwealth Bureau of Horticulture and Plantation Crops, East Malling, 131 (1953)
   The standard work on methods of experimentation with perennial plantation crops. Written by a statistician with an exceptional knowledge of field methods.

3. Harvey, P. N., *The Practice of Arable Crop Experimentation*, Norfolk Agricultural Station, Norwich, 79 (1952)
   The most comprehensive book published in England on field methods. Very good and clear but a little out of date now — and hard to find.

4. Hauser, G. F., *A Standard Guide to Soil Fertility Investigations on Farmer's Fields*, Soils Bulletin No. 11, Food and Agriculture Organisation (FAO), 71 (1970)
   Written mainly for the man who has to organise scattered experiments in a (tropical) developing country. Some of his statistical suggestions are a little unusual and as disturbing to me as mine probably are to him.

5. Unterstenhöfer, G., 'The Basic Principles of Crop Protection Field Trials', *Pflanzenschutz-Nachrichten 'Bayer'*, **16**, No. 3, 81 (1963)
Not strictly speaking a book, but a most compendious journal paper. Excellent on methods of assessing the efficacy of insecticides in killing insects; more specialised than those listed above.

## 2. BOOKS ON STATISTICS

Books on statistics (in the 'biometrical' sense) are legion and we all have our preferences (and prejudices). I suggest the following, all written by people very conscious of the nature and problems of field work.

6. Rayner, A. A., *A First Course in Biometry for Agriculture Students*, University of Natal Press, Pietermaritzburg, 626 (1969)
A first-class book, in simple English, with many examples worked out and explained in detail.

7. Finney, D. J., *An Introduction to Statistical Science in Agriculture*, Oliver and Boyd, Edinburgh, 179 (1953)
Much shorter but also an excellent book for anyone involved with field experiments.

8. Yates, F., *The Design and Analysis of Factorial Experiments*, Technical Communication No. 35, Commonwealth Bureau of Soils, Harpenden, 95 (1937, reprinted 1958)
Yates inherited from Fisher the habit of writing succinct accurate prose about statistics; this makes his papers (and this Communication) formidable reading but, like the Bible, very good to return to in times of difficulty.

9. Snedecor, G. W. and Cochran, W. G., *Statistical Methods*, Iowa State University Press, Ames, Iowa, 6th edition, 593 (1967)
A well-tried and readable book dealing with all statistical methods needed by a field experimenter (and more besides). This has grown out of the first edition (by Snedecor alone; same title) published in 1937.

10. Cox, D. R., *Planning of Experiments*, Wiley, New York, 308 (1958)
A shorter book, mainly on design of experiments but with some good examples of graphs used to summarise results.

11. Kendall, M. G., *A Course in the Geometry of n Dimensions,* Griffin, London 63 (1961)
Read this (or at least dip into it) if you found my chapter on *n*-dimensional geometry intelligible. I believe many statisticians think occasionally in the geometrical mode but few, other than Kendall, have bothered to write down their thoughts. Strange, though, that the book has no diagrams!

## 3. BOOK OF TABLES

12. Fisher, R. A. and Yates, F., *Statistical Tables for Biological, Agricultural and Medical Research*, Oliver and Boyd, Edinburgh, 6th edition, 146 (1963)
Contains tables of t, F, etc.; also squares, square roots, logarithms (natural and to base 10) and random numbers. Fisher and Yates give catalogues of designs – latin squares, balanced incomplete block designs, etc.

## 4. REFERENCES

13. Anon., 'The Use of a Mine Detector for Locating Turf Plot Marks', *J. Sports Turf Res. Inst.*, **8**, No. 29, 287 (1953)

14. Bartlett, M. S., 'Some Examples of Statistical Methods of Research in Agriculture and Applied Biology', *J. R. statist. Soc. Supplement*, **4**, No. 2, 137 (1937)

15. Beaven, E. S., 'Trials of New Varieties of Cereals', *J. Minist. Agric.* **29**, Part 1, 337, Part 2, 436 (1922)

16. Biometrics (Special issue on the analysis of covariance) *Biometrics*, **13**, No. 3 (1957)

17. Butters, B., 'Some Practical Considerations in the Conduct of Field Trials with Robusta Coffee', *J. hort. Sci.*, **39**, No. 1, 24 (1964)

18. Cleaver, T. J., Greenwood, D. J. and Wood, J. T., 'Systematically Arranged Fertiliser Experiments', *J. hort. Sci.*, **45**, No. 4, 457 (1970)

19. Collis-George, N. and Davey, B. G., 'The Doubtful Utility of Present-day Field Experimentation and other Determinations Involving Soil–Plant Interaction', *Soils Fertil*, **23**, No. 5, 307 (1960)

20. Crowther, E. M., 'The Technique of Modern Field Experiments', *J. R. Agric. Soc.*, **97**, 54 (1936)

21. Crowther, E. M. and Yates, F., 'Fertilizer Policy in War-time: the Fertilizer Requirements of Arable Crops', *Emp. J. exp. Agric.*, **9**, No. 34, 77 (1941)

22. Dyke, G. V. and Meyler, S., 'Indoor Experiments with a Combine Harvester', *Expl. Husb.*, No. 1, 63 (1956)

23. Dyke, G. V., 'Restricted Randomization for Blocks of Sixteen Plots', *J. agric. Sci. Camb.*, **62**, No. 2, 215 (1964)

24. Dyke, G. V., 'Field Experiments and Increases in Yield of Crops', *J. nat. Inst. agric. Bot.*, **11**, No. 2, 329 (1968)

25. Dyke, G. V., 'Experiments to Compare Combine Harvesters', *Expl Husb.*, No. 23, 10 (1973)

26. Engledow, F. L. and Yule, G. U., 'The Principles and Practice of Yield Trials', *Emp. Cott. Grow. Rev.*, **3**, Part 1, 112, Part 2, 235 (1926)

27. Fisher, R. A., 'The Influence of Rainfall on the Yield of Wheat at Rothamsted', *Phil. Trans. R. Soc.* (B), **213**, 89 (1924)

28. Fisher, R. A., 'The Arrangement of Field Experiments', *J. Minist. Agric.*, **33**, No. 6, 503 (1926)

29. French, B. K., Long, I. F. and Penman, H. L., 'Water Use by Farm Crops: I. Test of the Neutron Meter on Barley, Beans and Sugar Beet, 1970; II. Spring Wheat, Barley, Potatoes (1969); Potatoes, Beans, Kale (1968)', *Rothamsted Experimental Station, Report for 1972*, Part 2, 5 and 43 (1973)

30. Gasser, J. K. R., '*The Use of Fertilizers in Increasing Rice Yields in British Guiana and their Role in the Nutrition of the Rice Plant under Waterlogged Conditions*', Report on an investigation carried out under Colonial Development and Welfare Scheme D 1977, Department of Agriculture, British Guiana (1956)

31. Grundy, P. M. and Healy, M. J. R., 'Restricted Randomization and quasi-Latin Squares', *J. R. statist. Soc. (B)*, **12**, No. 2, 286 (1950)

32. Healy, M. J. R., 'Latin Rectangle Designs for $2^n$ Factorial Experiments on 32 Plots', *J. agric. Sci., Camb.*, **41**, No. 4, 315 (1952)

33. Healy, M. J. R., 'Multiple Regression with a Singular Matrix', *Appl. Statist.*, **17**, No. 2, 110 (1968)

34. Horne, B., 'Leys and Soil Fertility – Part I. Crop Production (First 12 Years)', *Expl Husb.*, No. 23, 86 (1973)

35. Jameson, H. R. and Mansbridge, B. E., 'Marking Out the Sites of Long-term Field Experiments', *Expl Husb.*, No. 10, 112 (1964)

36. Jenkyn, J. F. and Bainbridge, A., 'Problems of Designing Field Experiments with Air-dispersed Pathogens', *Rothamsted Experimental Station, Report for 1972*, Part 1, 133 (1973)

37. Kendall, M. G., *The Advanced Theory of Statistics*, Vol. 2, Griffin, London, 521 (1946) (better on this subject than the later three-volume version by Kendall and Stuart)

38. Mark, 'Gospel', Chapter 4. verses 3–9 (*c.*65)

39. Mercer, W. B. and Hall, A. D., 'The Experimental Error of Field Trials', *J. Agric. Sci., Camb.*, **4**, No. 2, 107 (1911)

40. Moffatt, J. R., 'Agricultural Methods Adopted in the Rothamsted Classical and Modern Field Experiments', *Emp. J. exp. Agric.*, **7**, No. 25, 251 (1939)

41. Moffatt, J. R., 'Cultivation Weedkiller Experiment -- Rothamsted and Woburn Report for Years 1961–65', *Rothamsted Experimental Station, Report for 1965*, 221 (1966)

42. Patterson, H. D., 'Theory of Cyclic Rotation Experiments', *J. R. statist. Soc. (B)*, **26**, No. 1, 1 (1964)

43. Patterson, H. D., 'The Factorial Combination of Treatments in Rotation Experiments', *J. agric. Sci., Camb.*, **65**, No. 2, 171 (1965)

44. Penman, H. L., *The Calculation of Irrigation Need*, Technical Bulletin No. 4, Ministry of Agriculture, Fisheries and Food, 37 (1954) 7th impression (1965)

45. Rothamsted Experimental Station, 'The Technique of Field Experiments', *Rothamsted Confs*, 13 (1931)

46. Rothamsted Experimental Station, *Details of the Classical and Long-term Experiments up to 1967*, Rothamsted Experimental Station, Harpenden, 128 (1970)

47. 'Student', 'Mathematics and Agronomy', *J. Am. Soc. Agron.*, **18**, No. 8, 703 (1926)

48. Widdowson, F. V., 'Results from Experiments with Wheat and Barley Measuring the Effect of Paths on Yield', *Expl Husb.*, No. 23, 16 (1973)

49. Wilson, J. C., 'A Spraying Machine for Small Plot Experiments', *Expl Husb.*, No. 21, 25 (1972)

50. Yates, F., 'Incomplete Latin Squares', *J. Agric. Sci., Camb.*, **26**, No. 2, 301 (1936)

51. Yates, F., *Sampling Methods for Censuses and Surveys*, 3rd edition, Griffin, London, 440 (1965)

# INDEX

Accuracy, loss of when degrees of freedom are lost, 7, 77
Additivity, additive model, 113, 153
Analysis of covariance, 144
Analysis (of variance), 114
  in relation to design, 14
  of observations (not yields), 153
Aphids, 77, 153
Application of basal manures, sprays, etc., 41
Area harvested
  adjustment for sampling, 56
  defined, 20
  choice of, 45
  marking out, 46
Arrows, surveyors', 37
Athens, 128
Auger, 37

Baby (and bathwater), 78
Background information, 81
Balance
  direct-recording, 48, 167
  spring, 48
Barley, Rothamsted Classical experiment on, 35
Basal manures, sprays, etc., 41
Base-line, 33
Basins, for irrigation, 28
Beans, 29, 153

Bias
  in sampling, 52
  in estimated coefficients of regression, 56, 88, 154
Binder, 64, 86
'Biological yield', 89
Blank-row technique, 21
'Blind' harvesting, 20
Blocks ignoring treatments, 126
Boots, damage by, 27
Broadbalk, 60, 64
Broadcast crops, compared with drilled, 46
Bunds, for irrigation, 30

Calculating machines (hand and electric), 170
Calculations, statistical, practical considerations, 169
Calibration (of seed-drills, etc), 40
Calories, 161
Can, watering, 32
Carbon fixation, 158
Carbon paper, 49
Chain, surveyors', 37
Chemical analyses, 76
Chessboard method, 94
Choosing
  of treatments, 5
  of design, 9
Classical experiments, 59

205